Palgrave Studies in Law, Neuroscience, and Human Behavior

Series Editors
Marc Jonathan Blitz
Oklahoma City University School of Law
Oklahoma City, USA

Jan Christoph Bublitz
University of Hamburg
Hamburg, Germany

Jane Campbell Moriarty
Duquesne University School of Law
Pittsburgh, Pennsylvania, USA

Neuroscience is drawing increasing attention from lawyers, judges, and policy-makers because it both illuminates and questions the myriad assumptions that law makes about human thought and behavior. Additionally, the technologies used in neuroscience may provide lawyers with new forms of evidence that arguably require regulation. Thus, both the technology and applications of neuroscience involve serious questions implicating the fields of ethics, law, science, and policy. Simultaneously, developments in empirical psychology are shedding scientific light on the patterns of human thought and behavior that are implicated in the legal system. The Palgrave Series on Law, Neuroscience, and Human Behavior provides a platform for these emerging areas of scholarship.

More information about this series at
http://www.springer.com/series/15605

Marc Jonathan Blitz

Searching Minds by Scanning Brains

Neuroscience Technology and Constitutional Privacy Protection

Marc Jonathan Blitz
Oklahoma City University School of Law
Oklahoma City, USA

Palgrave Studies in Law, Neuroscience, and Human Behavior
ISBN 978-3-319-50003-4   ISBN 978-3-319-50004-1 (eBook)
DOI 10.1007/978-3-319-50004-1

Library of Congress Control Number: 2016963257

Cover illustration: Modern building window © saulgranda/Getty

Printed on acid-free paper

This Palgrave Macmillan imprint is published by Springer Nature
The registered company is Springer International Publishing AG
The registered company address is: Gewerbestrasse 11, 6330 Cham, Switzerland

CONTENTS

CHAPTER 1

Introduction

Abstract Modern technology helps neuroscientists gain a better understanding of how brain activity generates our mental experience – but it simultaneously threatens to undermine the privacy of that experience. In a number of impressive recent experiments, scientists have used fMRI or other brain-scan technology to infer the content of words or images a subject is imagining, or memories she is recalling. In future years, such technology could conceivably allow law enforcement or other government officials to uncover thoughts, feelings, or memories a person is unwilling to share. This chapter raises the question of how we should assess this threat to mental privacy – and what response (if any) US constitutional law can offer to this concern. These questions are the focus not only of this introductory chapter, but of this book's subsequent explorations of US constitutional doctrine. This chapter emphasizes that – in elaborating this doctrine and how it applies to neuroimaging – courts and other legal actors should focus not only on the extent to which specific uses of brain-scan technologies affect the privacy accorded to particular mental states, but the extent to which they might undermine individuals' more general sense that their unshared mental life is shielded from external monitoring.

© The Author(s) 2017   1
M.J. Blitz, *Searching Minds by Scanning Brains*,
Palgrave Studies in Law, Neuroscience, and Human Behavior,
DOI 10.1007/978-3-319-50004-1_1

Keywords Brain scan · Mind reading · Neuroimaging · fMRI · EEG · functional Magnetic Resonance Imaging · Privacy · Constitutional rights · Fourth Amendment · Fifth Amendment · First Amendment · Searches and seizures · Self-incrimination · Freedom of thought · Freedom of speech · Intellectual privacy

The most private of all human experiences are unexpressed thoughts and feelings. Until they are translated into verbal expression or body language, such mental experiences are inaccessible to everyone but the person who is doing the thinking or the feeling. This has been apparent to the community of scientists who research our inner lives. "Dreams," as one psychologist notes, "cannot be observed by anyone but the dreamer while they are happening" (Domhoff 2003, 39). Researchers can thus learn about a dream's content only when the dreamer emerges from sleep, and tells what she can remember of her imagined travels. Those who study waking consciousness face a similar predicament: Although someone who is awake – unlike a sleeping subject – can describe her feelings and thoughts as they arise, this description is necessarily an incomplete translation. For better or worse, a person's internal images are – as one recent study of consciousness states – "available directly only to the owner of the mind in which they occur" (Damasio 2010, 70). They are, in the words of another, confined to a "private theater" or sorts, a place where each self is not only the creator of an internal mental script but also its only audience (Edelman & Tononi 2010, 20).

What is a barrier to science, however, is a boon to privacy. The inherently private nature of our mental lives may sometimes stymie the efforts of psychologists to understand our minds, but it also shields our internal thoughts from other observers who we very much want to keep out. Indeed, it provides each person the only space in which a person can be sure that his secret plans, goals, or fantasies will remain secret. Locked drawers can be forced open. Electronic storage centers can be hacked. But our inner thoughts will, by their very nature, remain hidden until we reveal them.

This, in any case, is the state of affairs with which we are familiar – and which has provided the background for our existing privacy laws. It is not, however, a state of affairs that will necessarily survive the coming decades of the twenty-first century. There is because there are two important cracks in the protective wall that nature erects around our thought process, both of which might be dramatically widened by emerging technologies. While an outside observer cannot observe a person's thoughts or feelings, that observer might draw inferences about our inner thoughts based on two kinds of phenomena that can be observed: (1) Observations or records of our external behavior and (2) The physical correlates or causes of mental states in biological activity in the brain.

The first of these is more familiar, to lawyers and lay people alike: While we could conceivably generate and develop an idea entirely inside of our own minds, that is usually not how we go about expanding our understanding of the world, or exercising our creative and intellectual powers. Rather, we draw upon – and create records of our thinking in – the external world. We purchase and read books, search for information on the World Wide Web, or engage in conversation and collaboration with others. Such reading choices and Internet activities are records of our thinking processes. In fact, they are often more than records: When a scientist scribbles down an equation on a notepad or a writer sets down story ideas in a journal, she is often not simply recording a thought fully generated in silent contemplation, but rather working to create that idea in the first place. The writing is an essential component of the thinking process, not simply a record of it (Clark, 2008, xxv).

Indeed, even when we are zealous enough about the privacy of our thoughts and plans to avoid writing them down – or revealing them in a series of Web searches – it is almost impossible to hide them entirely. We tend to reveal clues about our plans and interests, for example, in the stores and others locations we choose to visit, the people we choose to associate with, and the questions we ask of others as we seek information in our daily lives. Although it was, in the past, difficult for any one observer to piece all of these clues about our thinking, it becomes easier for government (as well as certain large corporations) to do so as more and and more of our interactions with the world are recorded – by the location tracking that occurs as we carry SmartPhones, by video cameras that record increasing amounts of what occurs public space, or in transaction data generated and stored each time we make a purchase, or have an event ticket, personal identification, or license plate scanned as we enter a monitored area. Courts applying constitutional law have already grappled with how to understand these technological threats to our mental privacy.

The other breach in the protection that nature provides for our mental privacy is less familiar to courts, and less pervasive and familiar a part of everyday life: With emerging forms of technology, scientists might infer our mental states from the biological activity that generates these thoughts. Consider what scientists have recently been able to figure out about what is occurring in individuals' minds from scans of their brains. Using functional Magnetic Resonance Imaging (fMRI) to study brain activity, they have been able to tell what type of object a person is silently contemplating: One type of signature pattern of brain cell activation arises when someone thinks of a hammer, another when someone thinks of a house (Shinkareva et al. 2008). Brain scans have likewise revealed which room within a "virtual house" a person is navigating through (Hassabis et al., 2009). They have revealed whether a particular environment was new or familiar to the viewer (Smith 2013). They have revealed which clip in a series of film clips a person was remembering at a given time (Chadwick et al. 2010). And, in some of the most impressive demonstrations of this technology's potential, they have allowed scientists to reconstruct – entirely from fMRI readings of the brain activity in an individual's visual cortex – a rough reproduction of an image she was viewing or a video she was watching (Nishimoto et al. 2010). Brain scanning technology, in other words, has allowed scientists to take a set of perceptions from an individual's "private theater" – and rescreen them for outside observers. Such a technology might one day let dream researchers watch a person's dreams – or at least elements of them – on a video screen, instead of learning about them second-hand later from a subject's best (but often somewhat unsuccessful efforts) to remember his imagined adventures and translate them into words. Indeed, such a primitive fMRI-based "dream recorder" was built by a team of researchers in Japan allowing them to determine (with 60% accuracy) which objects the subjects of the experiments had reported remembering from their dreams (Stromberg 2013).

The fMRI scans are not the only technology scientists have used to uncover the physical correlates in the brain of different cognitive tasks or feelings. Unlike fMRI technology, which can only be used when individuals lie in a massive cylindrical scanner inside a laboratory, functional near infrared imaging (fNIR) produces similar maps of brain activity, but does so by shining specific wavelengths of near infrared light into an individual's cortex from a portable head set (Ayaz et al. 2011). Electroencephalography (or "EEG") devices are also cheaper and more easily used than are fMRI machines: EEG has long been used to measure the rhythms of electrical activity ("brain waves") that occur as neurons generate electrochemical signals throughout the brain (Marcuse et al. 2016, 1–10, 12). They typically generate such measurements from multiple electrodes placed over a person's scalp (Marcuse et al. 2016, 1–10) but makers of video games and biofeedback devices have recently marketed headbands and helmets that allow individuals to transform brain waves into a form that can be viewed on a computer, or perhaps used as a system to control video game play (Childers 2013).

Certainly, at present, none of this technology – which I will refer to under the heading of "neuroimaging" – currently allows scientists, or anybody else, to engage in the kind of detailed mind reading one finds in science fiction and fantasy. Individuals cannot, as in Philip K. Dick's story, *Ubik*, steal commercially valuable data from each other's minds (Dick 2007). Existing neuroimaging is nothing like the magical "pensieve" device that characters in Harry Potter use to immerse themselves in each others' vivid three-dimensional memories (Rowling 1998). Nor is it like the fictional brain tapping that the heroes of the movie, *Inception* (2010) and the villains of the television show, *The Prisoner*, use to enter and pry secrets from one another's dreams (Carraze & Oswald 1996). The modern-day uses of neuroimaging – to divine mental content – focus on telling whether someone is lying or not, or whether someone recognizes an image or other stimulus. In such cases, it is the researcher who provides much of the content (in the form of a question to be answered, or a word or picture to react to), and the brain under observation then provides some indication of whether there is deception in the person's behavior, or that what they are seeing is familiar. Even these modest forms of "brain-based mind reading" (to borrow a term used by Francis Shen) (Shen 2013, 710) are in early stages of development, and not yet admissible as evidence of dishonesty, for example, in US courts. Nor are they yet in use by US law enforcement. Of course, that may change as the technology develops – and the examples of fMRI experiments that reveal more about mental content, like a word or concept that someone is concentrating on, or the imagery they are looking at in their mind's eye, provides hints of what more powerful neuroimaging may be able to do in the future.

My focus in this book is on the second of these two threats to our mental privacy: the possibility that neuroimaging will reveal unexpressed thoughts that we have succeeded in keeping out of Internet records, and hidden from government or other video cameras, or other surveillance technologies. How should courts, lawmakers, and legal scholars address this technology? First, how should they think about the threat to privacy it presents? Should they view it as a privacy concern *only* if and when it matures into something closer to the mind reading we currently find only in science fiction – something that might allow officials to look and listen into the private scenes and dialogues of unexpressed thoughts? Or should constitutional limits and other privacy laws block even the more limited and fragmentary glimpses that fMRI, fNIR and EEG technology might already provide into our mental operations? Do brain scans become a significant privacy problem only when they reveal substantially more about our thoughts and intentions than what may *already* be revealed by other evidence, like a journal entry or an illuminating conversation or interview? Or is there something inherently disturbing about officials – or other onlookers – observing, and drawing inferences from, the brain itself as we generate a thought or recall a memory, even if they have other, more traditional ways of learning of that thought's contents? And in answering such questions about the threat that neuroscience technology can create to our mental privacy, it is worth looking at how courts (and scholars) have addressed the more familiar ways that government can threaten such privacy: If government-imposed neuroimaging does raise constitutional privacy problems, are these problems of the same type as those already raised by government surveillance that infers our thoughts from external records or external behaviors? Or does neuroimaging merit its own constitutional analysis? As I explain more fully in this book, these are not simple questions. While it is easy to see what is horrifying about an Orwellian dystopia in which all our thoughts and feelings are on full display, it is less clear if and when neuroimaging technology should bother us when it reveals far less information far less frequently and also unclear whether the problems it does raise require new approaches to applying relevant constitutional principles, or straightforward adaptions of approaches courts already use to deal with other technologies.

Moreover, however we answer these high-level questions about the privacy implications of neuroimaging technology, these answers will not by themselves, tell us how the law should react to such technology. The law's response depends, to a large extent, not only on whether neuroimaging does or does not raise significant privacy concerns (and, if it does raise concerns, at what point this occurs), but also on how and when a legal system provides protection for individual privacy. My focus in this book is on the privacy protections of the US constitutional system – and specifically the protections found in three constitutional safeguards against certain exercise of government power: The Fourth Amendment's protection against unreasonable searches and seizures by governmental authorities, and especially police searches of private environments; The Fifth Amendment privilege against self-incrimination, which protects a criminal defendant against being forced to testify against himself; and the First Amendment's freedom of speech.

This last constitutional safeguard is not often thought of as a kind of privacy protection. It usually protects protestors and others as they draw attention to themselves and their views, not as they hide from the world. Perhaps, for this reason, the First Amendment has received relatively little attention in legal thinkers' explorations of what legal protections may constrain the use of neuroimaging.

However, the First Amendment also provides some assurance that individuals in the United States can express themselves, and seek out new ideas, *anonymously* – free from monitoring of a kind that might constrain their seeking out, or endorsement, of dissenting or unorthodox views (McIntyre v. Ohio Elec. Comm'n, 514 U.S. 334, 341–343 1995). It also, the Supreme Court has said, protects US citizens' "freedom of mind," (Wooley v. Maynard, 430 U.S. 705, 714 1977) and it is possible that government undermines that freedom not only when it constrains or punishes such thinking, but also when it monitors it – letting government watchers access, and possibly punish, individuals' private intellectual heresies.

I argue in this short book that this First Amendment emphasis on "freedom of mind" is actually crucial for thinking about how the more familiar constitutional privacy protections – in the Fifth and Fourth Amendments – should apply to neuroimaging. In short, I argue, the latter amendments should be seen not simply as providing robust protection for privacy, but must often provide an especially strong layer of protection for what some scholars have called "intellectual privacy." "Intellectual privacy," wrote Neil Richards, is "protection from surveillance or interference when we are engaged in the processes of generating ideas – thinking, reading, and speaking with confidants before our ideas are ready for public consumption" (Richards 2015, 5). Such freedom, he argues, both derives for, and is a crucial support for, freedom of speech. Thus, where a government search would not merely intrude into citizens' private lives, but intrude into the privacy of their *mental* lives, the Constitution should demand the same kind of heightened scrutiny (and wariness) of such intrusion into thought as the court shows for threats against their speech – and this, I argue, is true even though it is traditionally the Fourth Amendment, not the First, that courts turn to in order to judge the constitutionality of a government search.

This should be true, I argue, even when a specific government intrusion into our mental lives does not, by itself, seem to threaten significant damage to mental privacy in a specific instance. Not every such intrusion will be Orwellian. Sometimes, law enforcement's use of an fMRI or other neuroimaging device will do nothing more than gain access to a memory or other knowledge that a witness is, in any case, obliged to share (and convey honestly). And while law enforcement agents should perhaps be able to obtain such neuroimaging evidence, under careful judicial monitoring, that does not mean they should necessarily be able to compel someone to provide such evidence easily or frequently or as a first rather than a last resort. Government, after all, cannot generally force its way into a person's house without a warrant even if it promises to view only mundane details and avoid looking at private information there. As the Supreme Court has said, "[i]n the home . . . all details are intimate details, because the entire area is held safe from prying government eyes" (Kyllo v. United States, 533 U.S. 27, 37 2000). And it is not only in the home that the "[s]tate should not be a dominant presence," the Court said in a different case, but also in the realm "of thought, belief, [and] expression" (Lawrence v. Texas, 539 U.S. 558, 558 2003). In these realms too, it may be the case that "all details" are private details, in the sense that the state should not be permitted to monitor or control them without strong justification. What is important, in other words, is not the privacy accorded to particular mental states, but the presumptive privacy accorded to the entire sphere. Thinking about constitutional privacy protection through this lens requires thinking about Fourth Amendment and Fifth Amendment law in a somewhat different way from the way that scholars (and courts) frequently understand them.

Thus, even with uncertainty over where neuroimaging technology is headed in the coming years, and how constitutional provisions will (and should) evolve during the same period, there is benefit in asking how constitutional principles might (and should) apply to technologies that potentially give government a window of sorts into what is, and is long been, a deeply private realm of human experience. This book will do so in four parts: Chapter 2 introduces and frames the constitutional problems in more detail. After looking at some general questions about how law enforcement neuroimaging might – or might not – threaten our mental privacy, it provides a quick overview of the Fourth Amendment and Fifth Amendment questions legal scholars have already raised about neuroimaging, and considers why the First Amendment may also have importance. Chapter 3 then takes a brief break from this legal discussion to look a little more closely at existing neuroimaging technology, and imagine the future uses it might have for law enforcement and in the legal system. Chapter 4 then focuses specifically on the Fifth Amendment self-incrimination arguments that have been the core of the scholarly discussion about neuroimaging's constitutional implications. Chapter 5 then turns to the more complex issues of Fourth Amendment protection, and the First Amendment template that I will argue should, to some extent, guide it. As I explain here, the Fourth Amendment puzzles raised by neuroimaging are a little more complex than those raised by the Fifth Amendment self-incrimination. Legal thinkers pondering neuroimaging's implications do not face, in analyzing the Fourth Amendment, the clear constitutional fork-in-the-road that they face when dealing with the Fifth Amendment (trying to figure out if mental content extracted by mind machines is "testimonial" evidence, covered by the Fifth Amendment's protection against self-incrimination, or "physical" evidence that is not). Rather, they have in the Fourth Amendment, a doctrine that has clearer implications in some ways (because its privacy protection almost certainly extends to, and "covers" the kind of thing government does when it collects information of any sort from inside our bodies, including our brains). But it creates confusion in others (because the "protection" it offers individuals against searches can vary markedly from one type of search to another, often in ways that are hard to explain or make sense of). Finally, Chapter 5 will also examine a constitutional provision that has received far less attention in legal debates about neuroimaging but looms large when one is talking about other constitutional provisions: namely, questions of First Amendment freedom of thought and expression. It will argue that the First Amendment can and should shape the way we apply the Fourth Amendment to emerging technologies of neuroimaging.

CHAPTER 2

Constitutional Puzzles and (Neuro) Technological Changes

Abstract This chapter explains why neuroimaging raises constitutional puzzles, even where constitutional rules at first seem clear. The Fifth Amendment bars compelled self-incrimination and one might assume that would prevent police from circumventing this limit by obtaining evidence of mental states some other way. The Fourth Amendment would almost certain classify neuroimaging as a search, and thus subject it to constitutional limits. However, both of the implications of these provisions are unclear: They seem to leave police with plenty of room to gather physical evidence of various kinds – and there are certain respects in which neuroimaging evidence resembles such physical evidence (as the chapter illustrates with the help of a hypothetical crime investigation). The chapter points to a way ahead and also argues that while the First Amendment isn't generally considered a kind of privacy protection, its freedom of thought protection may be a key part of solving these puzzles.

Keywords Brain scan · Mind reading · Brain-mind distinction · Evidence · Police · Law enforcement · Video evidence · Locke · Mill · Liberalism · Neuroimaging · Extended mind · Chalmers · Clark · Fourth Amendment · Fifth Amendment · First Amendment · Searches and seizures · Warrants · Warrantless searches · Self-incrimination · Freedom of thought · Privacy · Autonomy · Intellectual privacy · Internet privacy

© The Author(s) 2017
M.J. Blitz, *Searching Minds by Scanning Brains*,
Palgrave Studies in Law, Neuroscience, and Human Behavior,
DOI 10.1007/978-3-319-50004-1_2

How should courts and lawmakers address neuroimaging technology? A common reaction of writers covering it is to characterize it as a dire threat that merits strict regulation. One commentator, for example, warns that such technology brings us closer to an Orwellian society where "thought police" ferret out "thought crimes" (Federspiel 2008, 865–866). Even those who believe that such authoritarian uses are not likely to become commonplace in US society might still feel that the government, as a general matter, has little business observing thoughts and feelings someone has chosen not to share. To the extent fMRI or other brain-based mind-reading technologies widen a crack in the wall nature erects around our thought processes, one might argue that the law should seal it up again. In fact, one might assume that at least in the United States, certain constitutional provisions – such as the Fourth and Fifth Amendments of the US constitution – already do so.

As noted in the Introduction, this book argues for a robust right to mental privacy. But to do so, it has to contend with, and address, certain powerful concerns that seem to cut the other way. First, judges and citizens are understandably wary of approaches to individual rights that they believe would deal a grave blow to the safety and justice interests that government is charged with protecting. As I explain more fully later, the US Supreme Court and other courts have rejected approaches to Fourth Amendment search and seizure rights that they believed would severely hamper the ability of police to investigate crime. As one federal appeal court has stressed, the Fourth Amendment's proscription against "unreasonable search and seizure" cannot be read to deprive police of modern crime-fighting technology: It "cannot sensibly be read to mean that police shall be no more efficient in the twenty-first century than they were in the eighteenth" (United States v. Garcia, 474 F.3d 994, 998 (7th Cir. 2007)). In past cases, the Supreme Court, has insisted that the Fourth Amendment's proscription should not readily be interpreted to bar police from using investigative methods that are crucial (in the Court's view) for police and government to begin to build a case against sophisticated criminal operations: Where those conducting "organized criminal activities" use intimidation or other methods to prevent their victims from reporting such crimes to police, the Fourth Amendment should not prevent police from using undercover informants (Lewis v. United States 1966, 210). Where drug-growing operations cloak themselves inside the strong privacy the US society affords the home and its surrounding "curtilage," police – the Court insisted – must be left with *some* way to detect such operations from outside, for example, by following up on anonymous tips with "fly overs" by planes or helicopters (Ciraolo v. California 1986, 213). Even First Amendment free speech rights, as strong as they are in the American system, have been interpreted to leave room for government's ability to thwart violence, fraud, or other threats to public safety.

Similar concerns might well extend to mental privacy rights – in the First, Fourth, or Fifth Amendment context – that prevent police and prosecutors from using neuroimaging evidence. As much as individuals may want to keep certain memories or other thoughts private, there are certain circumstances where society may have a strong claim to them. As Adam Kolber writes, while individuals should normally have extensive freedom to shape their own mental experience and memories, whether in old-fashioned ways or though new technologies, it is also true that "our memories are not entirely our own . . . and we ought not have unfettered control over them" (Kolber 2008, 145). A "witness to a horrific crime," for example, he observes, might be obliged to preserve and share a memory of the event at trial, so that justice can be done (Kolber 2008, 145). Kolber's focus is on when individuals might be legally obligated to preserve memories they are technologically capable of erasing, but the point applies also to law enforcement's ability to access memories that certain witnesses refuse to share, or find themselves unable to articulate. When crimes, or the events leading up to them, occur far beyond the view of any video-camera or other recording technology, law enforcement will predictably benefit from instead obtaining whatever evidence the past leaves in the memories of participants and witnesses. They already try to do so by compelling testimony from such witnesses (although the Fifth Amendment bars them from doing so when the witness is the criminal defendant). But law enforcement officials and trial fact-finders do not simply take such narratives at face value. They scrutinize them in light of other, sometimes more powerful, evidence they might find in documents, video-recordings, DNA evidence, or other clues about what unfolded. So, it is not surprising that they may wish to add to this list of alternatives to spoken testimony, whatever clues the past has left in the brain activity of participants, and specifically in the biology underlying their memory of events. Indeed, as Rita Carter emphasizes, neuroimaging may be better than the alternative sources of evidence: "Conscious eyewitness recall is terrible, and mistaken recognition is responsible for more convictions of the innocent than all other factors combined. Most people can detect lying at little better than chance. And if information must be extracted, surely brain scanning is more humane than torture?" (Carter 2015, 145). Thus, while individual privacy protection should sometimes expand to keep up with new technologies, courts are unlikely to let it expand to the point that it shuts police out of all sources of evidence that they, and trial fact finders, need to address criminal activity.

To be sure how well neuroimaging can ultimately play such a role depends on how well it can overcome some potentially significant problems in its value as evidence – problems that may prevent it being admissible (let alone powerful) in American courts. Other writers – among them neuroscientists, psychologists and legal scholars – have emphasized at least three such problems.

First, as I have noted earlier, it is not yet clear exactly what kinds of inferences about our memories and mental states neuroimaging will allow scientists (and others cooperating with them) to draw in future. As Francis Shen writes, while neuroimaging devices may well continue to improve the inferences that scientists can make, "the complexity of the mind-brain relationship will prevent the government from using the brain data to reliably read individual minds" in the manner described in science fiction accounts (Shen 2013, 712–713).

Second, even if neuroimaging can allow scientists to infer certain mental states (such as dishonesty) in a controlled laboratory setting, this does not mean that it can do so with equal effectiveness in the very different setting of a police interview, or other law enforcement investigation environment. As Tenielle Brown and Emily Murphy write, "the behavior being solicited in response to the task" in a neuroimaging experiment "is usually so isolated that the results are difficult to generalize to other real-world functions" (Brown and Murphy 2010, 1143). A recent Macarthur Foundation Report on fMRI lie detection emphasizes another reason that such a translation from experimental fMRI inferences to real-world mind reading may be problematic when police or others seek to generate court-ready evidence of honesty or dishonesty: "Real-world fMRI lie detection focuses on events or facts that are likely to have occurred months or even years before, are deeply relevant to the subject, and have serious consequences. Little is known about whether real-world and experimental conditions yield similar results" (Wagner, et al. 2016, 4). A variation of this problem arises because, as Barbara Sahakian and Julia Gotwald write, "most neuroimaging studies are based on groups rather than on individuals" and this "means that we can rarely draw conclusions about the individual" (Sahakian and Gotwald 2017, 9). As Brown and Murphy write, use of group data to make an inference about an individual "may be highly problematic in a forensic and individualized legal context" (Brown and Murphy 2010, 1150).

Third and finally, even where fMRI or other neuroimaging technology does allow law enforcement to accurately infer that a person has a particular memory, this does not mean that the remembered event actually occurred in the way the person remembers it. Studies have shown that the memories individuals rely upon to make eyewitness identifications are often incorrect. In a series of experiments, Elizabeth Loftus has shown how experiments can create false memories in subjects, and how similar inaccuracies might arise in other, less controlled settings (Loftus and Ketchum 1994, 5). Others have noted that, unlike a video camera recording, memory is frequently revised each time it is recalled: As Jane Campbell Moriarty points out, human memory "does not record and recall information like a video recorder, but layers memory over memory, changes, loses, restructures, and adapts to continual addition of new information. Every time a memory is recalled, it is altered" (Moriarty 2009, 752). Unless those who use neuroimaging find a way not only to "read" such memories, but also to distinguish accurate from inaccurate memories, such evidence may be of limited value. As Jennifer Bard argues, "everything known about the flaws of memory for oral statements and witnessed events applies just as much to information extracted via neuroimaging" (Bard 2016, 352).

It may be that, even where such challenges are not completely overcome, neuroimaging technology will make enough progress in meeting them to compare favorably to alternative methods of gathering evidence. As Frederick Schauer points out, what matters is not whether neuroimaging provides perfect evidence of the events at the core of a trial, but rather whether it is better than the alternatives (such as questioning, and observation, of witnesses on the stand) (Schauer 2010, 1213). (Some observers are also understandably worried that even if neuroimaging evidence is comparable in value to witness testimony, its scientific source will make it seem more persuasive to jurors and judges than it really is.)

All of these issues are extremely important in determining when neuroimaging evidence will be reliable enough to meet the "Daubert" standard that governs when expert testimony is admissible in federal courts – or the "Frye" test used by some states' evidence rules. They are also important in any full consideration of whether government law enforcement interest in using such evidence can be strong enough to justify the sacrifice of privacy they entail: The lower the accuracy and reliability of the neuroimaging evidence, the less benefit society gets in return for whatever privacy is sacrificed when neuroimaging evidence is used. Still, for purposes of simplifying the inquiry here, I will focus – like some other explorations of neuroimaging's constitutional implications – on a more straightforward (but still difficult) question:Even assuming that all of the above challenges can be overcome, and neuroimaging can allow inferences about mental states that are admissible and useful for the justice system, when should constitutional privacy rules continue to present a hurdle to law enforcement use of such technology?

Even if we assume that neuroimaging can – or will one day be able to – make a significant contribution to criminal investigations and trials, it may be the case that, in some instances, a free society has to sacrifice a crime investigation's (or trial's) prospects of success in order to preserve its freedom. And, one can quarrel with specific Supreme Court predictions, like those regarding under-cover informants and aerial surveillance, about the damage that constitutional constraints on those methods would do to police investigations. But, as a general matter, the constitutional order one finds in the United States, and, in a somewhat different form, in other liberal democracies, is committed not only to individual autonomy, but also to protecting the state's capacity to protect lives and liberties from crime and other security threats. Consider John Locke's liberalism, for example, which provides the template for key elements of the American system of rights. On Locke's view of government's proper role, it is up to each individual what religious beliefs to espouse – because such beliefs are a matter of the "inward persuasion of the mind," over which the "outward force of the state" has no appropriate authority. But government, far from being powerless to restrict individual behavior in Locke's view, must have power to protect "life, liberty, health, and indolency of body," as well as property rights. John Stuart Mill similarly insisted government can have no legitimate authority over those aspects of an individual's life that "merely concer[n] himself," which affect only "his own body and mind." But he does not exclude the state from holding the individual "accountable" "for such actions as are prejudicial to the interests of others." Not all constitutional rights are necessarily aimed at marking this boundary line between a realm of individual autonomy and that of legitimate state power. But this boundary line is a key part of the background for much of the constitution's liberty and privacy protection – and helps us begin to place the mental privacy that might be threatened by emerging forms of neuroimaging. In short, when the state forces neuroimaging on an individual in order to gain access to mental content she wishes to keep private, we may, initially, be tempted to treat any such measure as intrusion into the realm Locke viewed as the "inward" realm of conscience or that which Mill viewed as the realm of an individual's purely self-regarding control over "mind and body," a realm where, in Mill's words, it is not the state or society but rather "individual" who "is sovereign." However, the writers I have cited above raise reasons to doubt this is always true: Where the information hidden in someone's memory is perhaps the only evidence available for solving a crime, or bringing its perpetrator to justice, then such mental content may well be content the public needs to protect individuals' lives or liberties, and a person's withholding such information could be "actions as are prejudicial to the interests of others."

I will ultimately argue that robust rights of mental privacy are compatible with these observations: Just as the government is, in most circumstances, shut out from our homes or bodies, even though it may – in exceptional circumstances – have a justifiable need to search there (when it can satisfy the conditions for a warrant), so it might be generally kept out of our mental lives except when it has extraordinarily good reason to enter this sphere. To be sure, there are some complexities we have to face to reach such a conclusion, or elaborate its meaning. First, constitutional privacy protection is, as noted earlier, not purely about marking this line between a realm of individual autonomy and of legitimate state power. According to many writers, for example, the Fifth Amendment's self-incrimination clause does not have this purpose at all: Its protection of criminal defendants from self-incrimination is about preventing state wrongs that have little to do with threats to the defendant's autonomy or privacy. Moreover, while Fourth and First Amendment law are, to at least, some extent, about keeping the state from interfering in spheres were individuals are left free (from state surveillance, or from interference in speech and thought), some of the doctrinal puzzles that neuroimaging raises in these areas (for example, whether and when it can be a "reasonable" search) may require going *beyond* simply marking a boundary line between spheres reserved for individual privacy and for more active state regulation, and require understanding when precedents allow such a boundary line to be crossed. Second, all uses of neuroimaging may not deserve the same analysis: Some may be less threatening to privacy than others, or more justifiable for other reasons.

Video Cameras, Parrots, and Brains

Let us elaborate a little about each of these points, beginning with the second. Police and prosecutors might well argue that even if state use of neuroimaging as a "mind reading" seems to be – and potentially is – a privacy-threatening technology, there are some forms of it that should be far less worrisome, and far less threatening to constitutional rights. Consider two variants of such an argument. First, they might stress that some uses of neuroimaging are – as Francis Shen suggests may be the case – not mind reading, but "brain reading." They reveal aspects of a person's brain activity. Such brain activity *might*, as noted in the introduction, be used to draw inferences about mental states. But, some might argue, such brain activity evidence could also be used to draw inferences about an individual's interactions with the world *without* drawing any conclusions about specific mental content. And, in this case, they might argue, that while our internal brain activity (like other bodily activities or conditions) may be private, it does not have the same claim to privacy as our unshared beliefs or feelings. Second, even where state-compelled neuroimaging concededly is used to draw inferences about mental content, some such content may less private than others, or otherwise more reasonable for the state to access.

Brain-Reading

Consider, first, the argument that use of neuroimaging for brain reading is possible, and more acceptable than use of it for mind reading. There is, of course, a close connection between the operation of the brain and the thoughts we have. A person's thoughts and feelings are somehow generated by brain activity, and distinctive thoughts and feelings appear to be gener-ated by distinctive patterns of brain activity. So if scientists can match the two – if they can figure out, for example, what pattern of brain activity tends to occur when a person is accurately remembering a particular living room setting – they'll only need to detect one side of this match to infer the presence of the other. For example, if the fMRI shows – in my head – the brain activity that scientists have previously associated with imagining that living room, they may be able to infer that if they see that brain activity in my head, that means I'm likely to be remembering the living room. Mind reading becomes, brain reading, in other words, only because scientists can create such matches, and use sophisticated computer algorithms to build "dictionaries" to let them (or more likely, a computer) figure out which mental states go with which brain activity patterns and vice-versa.

However, if government officials using neuroimaging are accused of enga-ging in mind-reading that objectionably intrudes into our mental privacy – there mind-brain distinction suggests one obvious response: They're not aiming to read anyone's mind, they might say, but rather simply to look at the brain – and see if it responds in the way one would expect it to respond if the person whose brain it is were in a particular location, or performed particular actions in the past. Imagine, for example, that police are investigat-ing a murder. The suspect they have arrested claims never to have met the victim, let alone been in her living room (where her body was found). That suspect may be lying, however, and if he is, police may uncover clues in the world that reveal what really happened – that he has met the victim (and may have killed her) and that he has been in her living room. Police may, for example, find a text message he has sent to her. Or a piece of the suspect's hair in the living room he claims he never entered. Or a trace of the victim's blood on his clothing. They also might find patterns of brain activity that he is unlikely to have if he never met the victim or never entered the living room – brain activity that would not be likely to arise in the suspect unless had seen and retained a memory of the victim's face, or been in, and remembered certain aspects of, her living room. Reading such brain activity, government officials might argue, is not the same as mind reading: Perhaps the suspect does not have any conscious memory of seeing the victim or being in her living room. It is not important, they might say, what he thinks or is aware of. What is important, they might say, is that his brain shows evidence of his having been there and seen the victim. This, they might say, is permissible brain reading, not invasive mind reading.

To see why this argument presents a plausible challenge to an absolute rule against neuroimaging, it is helpful to imagine another hypothetical and somewhat fanciful criminal investigation that elaborates upon the one above. Imagine that a New York city mansion is the scene of a robbery and murder. The woman who lives there is found strangled in her living room. Various items, including jewelry and a valuable painting have been stolen. And it appears as though she was coerced, prior to her killing, into transmitting money to a foreign bank account on her laptop computer.

Police have apprehended two suspects in the murder: Ozzie and Ivy. Ozzie brings them an unusual source of potential evidence. Ozzie has a long-term memory problem (caused by an injury). Because of this, he has a tiny video camera implanted his skull, so that it peers out from the surface of his head and constantly records what is in front of eyes. It then copies its footage to a small computer chip implanted in Ozzie's brain, where he can draw on its record of events to replace the now-damaged brain processes that, in other people, allow for the successful retention and recollection of long-term memory. Ivy has no such artificial memory aid. Her brain is injury-free and enables her to remember events from long ago. Indeed, she has an excellent long-term memory even for fine details.

Imagine police decide to take advantage of the video feed Ozzie uses to supplement his memory. They hire a skilled technician who has figured how to remotely copy the video footage from Ozzie's external micro-camera. In the event Ozzie has somehow taken steps to erase the footage they suspect will show Ozzie robbing the mansion and participating in the murder, they also ask the technician to develop a method for copying information from the computer chip that stores the video feed in Ozzie's head – and then translates it into signals Ozzie's brain can access and process. Would either the hacking of the micro-camera footage, or the computer chip, be mind reading of the sort that might be impermissible and, perhaps illegally, infringe Ozzie's privacy? The police might answer with a firm "no:" They have not forced Ozzie to share any of his own feelings, thoughts, and memories. Instead, they've obtained a record of the past from another source – namely, camera footage that may help to supply Ozzie with visual and auditory data that he uses in construction of his memory, but is not equivalent to his memory. It has an independent existence, such that police can access it without forcing Ozzie to share his private mental experience. Of course, the camera footage, much to Ozzie's disappointment and frustration, may reveal his participation in the crime. But it is not revealing his mental states. It is rather revealing images of the same sort police could find in an external video camera mounted within the mansion that Ozzie and Ivy have robbed.

But if police can permissibly access the micro-camera footage, and the transfer of it on the computer chip, might they also permissibly access any biological equivalents of this footage in Ivy's brain? Human memory, to be sure, operates very differently from a video camera. It is much more malleable: Memories can be transformed as they are recalled and the memory of an old event is refashioned in light of present experience. But to the extent the biology underlying memory does preserve – in *some* form – changes in neurons' connections with each other that allow a person to accurately recall certain aspects of her past experiences, might police look at this biological basis for memory generation – and see if they can use it to infer what a person has done? And might police and prosecutors argue that just as looking at Ozzie's video camera and chip (in their view) is quite different from commandeering and exploring Ozzie's own mind, so looking at Ivy's brain wiring is also different from commandeering and exploring her private mental experience?

This criminal justice hypothetical is adapted from another, less crime-oriented hypothetical offered by the philosophers, Andy Clark and David Chalmers. Clark and Chalmers similarly ask us to imagine two friends – one of whom, Otto, stores some of his memory on the outside of his brain, and the other of whom, Inge, stores her memory entirely on the inside of her brain. Like Ozzie in my example, Otto's brain no longer creates long-term memory effectively (in his case, because of Alzheimer's), so he writes information he needs to remember in a notebook that he carries everywhere and consults regularly. Inge generally needs no such notebook to remember information. So, if Otto and Inge agree to meet at the Museum of Modern Art, they will likely remember its address in different ways: While Inge will remember the Museum's address (after having visited in many times before) by simply remembering where it is, Otto will have to look it up in his notebook. Clark and Chalmer's point is that the set of mental acts we call "mind" is not performed solely by operations in the gray matter inside our heads. It is performed also with physical tools that extend outside our brain, and outside our bodies. While people may resist saying that Otto's notebook are part of machinery that allows him to generate "mind" or "memory," Clark and Chalmers argue that it has as just as strong a claim to that title as the pattern of neuronal connections in Inge's brain that enables her to recall the Museum address. This is an argument, for what they call the concept of an "extended mind."

Clark and Chalmers' analysis is interesting in part because it provides the basis for a challenge to the earlier-discussed police claim that, by using Ozzie's video record, they are steering clear of his mind and focusing only on physical evidence. If the notebook Otto constantly uses to remember events is a part of his mind and mental processes, this is surely true also of the camera and computer chip that Ozzie uses for the same purpose. Perhaps then police *do* engage in a kind of mind reading when they look at Ozzie's video feed, or for that matter, a notebook of Ozzie's – if he ever uses that in the same way as Otto does (perhaps as a back-up for the high-tech memory supplement in his camera-chip set up). As Clark and Chalmers argue, one possible consequence of their extended mind argument is that "[i]n some cases, interfering with someone's environment will have the same moral significance as interfering with their person" (Clark and Chalmers 2008, 232). Neil Levy likewise advocates an "ethical parity principle," holding that "[u]nless we can identify ethically relevant differences between internal and external interventions and alterations [in the mind], we ought to treat them on a par" (Levy 2007, 129–131). Perhaps in some cases, there should also be a legal or constitutional parity principle, such that when police help themselves to the video footage in Ozzie's camera-computer chip set up (from either the chip itself or the camera itself), they are doing something just as concerning as what they would be doing if they somehow extracted information about Ivy's memories from her brain.

But the parity principle could also conceivably cut the other way: If our intuitions tell us that it should be permissible for police to consult Ozzie's video-recording, why should it be impermissible for them to consult the natural equivalent? Would they really be reading Ivy's mind without her consent in the latter circumstance? Or would they be by-passing her conscious thought and feeling to simply find, in her neurons' connections with each other, information about her past actions that is no more inherently private than the record left by light and sound waves in Ozzie's video apparatus and connected computer chip?

Moreover, there is another point that one might raise against the claim that *either* police access to Ozzie's video-camera feed and computer chip *or* their access to Ivy's neuronal patterns would count as "mind reading" of any kind. That certain physiological or physical events (in the brain or an attached computer chip) underlie and generate mental states or processes does not make those *equivalent* to mental states or processes. Dennis Patterson and Michael Pardo have recently emphasized this difference between psychological activities such as "thinking or perceiving" and the "brain states" or "patterns of neural activity" that accompany it. Certain brain states or neural activity can be correlated with thoughts, beliefs, or other psychological activities, but they do not "constitute" such activities (Pardo & Patterson 2013, 11). It is thus, somewhat misleading to describe that the video camera or computer chip connected to Ozzie's brain (in my hypothetical above) or Otto's journal (in the Clark and Chalmer's hypothetical from which mine was adapted) as a part of Ozzie or Otto's "mind." They are rather physical sites where events occur that may generate, and be essential for, mental processes but are not equivalent to them. The same is true of the patterns of neuronal firing that arise when Ivy or Inge remember events or addresses.

To add to the police's argument before responding to it, we might add another fact to my earlier hypothetical: Imagine that the murder victim owned an African gray parrot. The parrot is found unharmed at the crime scene, and police notice that it frequently repeats the phrase "tell us where the jewelry is" and "tell us, or we'll kill you." They believe the parrot is repeating words it heard the killers say to the victim shortly before the murder. And they'd like to see if, when they use an fMRI to scan the parrot's brain, it reacts differently to a recording of Ozzie and Ivy's voices than to other voices – perhaps because it recognizes one of their voices as the one that voiced the words it is now repeating. (This example is not entirely fanciful: *The Washington Post* reported in June 2016 that police had discovered a parrot repeating words apparently said by a murder victim just before his killing (Holley, June 5, 2016)). Would such evidence of the parrot's brain activity be mind reading? The government may well argue it is nothing of the sort: No one knows what a parrot experiences, or whether it experiences anything at all. For the same reasons that the philosopher, Thomas Nagel, gave that we can never understand what it is like to be a bat (Nagel 1974), they might insist that they can't understand – and don't aim to understand – what it is like to be a bird. They are therefore interested not in the parrot's past or present internal point of view – but rather in what they *can* measure, which is whether parrot shows brain activity it is unlikely to show unless it heard Ozzie or Ivy nearby. And if such a defense against a charge of mind reading can work in the case of the African gray parrot, why not also in the case of a person?

It is true, of course, that people have constitutional rights – under the First, Fourth, and Fifth Amendments – that parrots have no claim to. We respect people's privacy and dignity – and insist government respect people's privacy and dignity – for reasons that courts and most citizens probably believe do not apply to animals. But such a different doesn't answer the argument here. The argument is not that Ozzie and Ivy lack moral or constitutional rights that they can raise against government coercion, or that their rights are equivalent to the (likely non-existent) rights of the parrot. It is rather that, if we want to understand what government would be doing in reading Ivy's brain activity, we can't simply assume that it involves reading her mind – any more than it is mind reading for police to read the brain activity of another biological creature. Police and prosecutors, of course, are likely to have a better sense of what goes on in Ivy's mind than in the mind of another creature: We assume other human beings have many of the same feelings and mental experiences that we do. But Ivy's internal mental experience, they might claim, is not what they are after when they put her in an fMRI scanner while investigating a crime. Rather, they are trying to establish the same thing they are trying to establish when they image the parrot's brain: Namely, to answer the question of whether there exists evidence, in her brain activity, or in that of the parrot, that certain patterns in the brain arise that would be extraordinarily unlikely to arise unless Ivy was in the victim's living room at some point in the past? These kinds of arguments complicate the task of thinking about whether and when neuroimaging raise privacy problems, or triggers constitutional privacy protections.

Less Private Mental States

There are, however, good reasons for at least some skepticism toward the above argument. Neuroimaging of brain activity that correlates with a person's having been in a certain place or seen certain events seems intuitively to invade that person's mental privacy – and there are reasons one might offer to justify this intuition. First, as noted above, while we may not know what a parrot experiences when it hears (and apparently remembers) certain voices, we do have a sense of what other people experience when they see and later remember a certain place or event. We know that, as a general matter, if Ivy's brain has presented a record of an event, and responds with kind of brain activity that correlated with past presence, this is generally only because Ivy has a memory of the event and recognizes it as familiar. As a consequence, even if police and prosecutors claim to be unconcerned with Ivy's internal experience of remembering her participation in a crime, or presence at the crime scene, government awareness of that internal experience may come packaged with the acquisition of information that connects her to that event or scene. Second, even if brain activity is not being used to infer mental states by police at the time they are conducting neuroimaging to gain information, if such records of brain activity could be used to infer mental activity at a later time, then privacy is threatened – just as it is when DNA obtained from a person solely to connect him with another DNA sample (at a crime scene) is later used to obtain information about his biological characteristics (Scherr 2013, 472–473).

Government officials may, however, offer other reasons to argue that certain kinds of compelled neuroimaging should not worry us: Even if neuroimaging is used to generate information about a person's mental states or mental processes, some of such information may not be as private as the kind of "mind reading" technology that is most worrisome. In Dov Fox's (2009) view, for example, it is of legal significance that currently available versions of brain imaging are "not capable of exposing the content of a subject's cognitive thoughts and propositional attitudes, such as normative judgments, religious convictions, and hopes or fears for the future" – and instead give government access only to "the less privileged sphere of sensory recall and perceptual recognition about a particular set of facts or the state of past events." Fox, to be sure, nonetheless argues that self-incrimination law should nonetheless protect an individual from even this kind of compelled neuroimaging because it should protect "a suspect's control over his thoughts from unwanted government access and use." (p. 797).

But government could conceivably build upon an observation such as Fox's to argue that in many contexts (if not in self-incrimination law), an individual should receive less protection against compelled neuroimaging when that neuroimaging *only* reveals memories of interactions with the outside world, and steers clear of revealing "normative judgments, religious convictions, and hopes or fears for the future," or other thoughts or feelings that might remain completely inaccessible to (real or hypothetical) outside observers in a way that memories of events are not. Consider again the dream experiences or other feelings I described in the introduction as "inaccessible to everyone but the person who is doing the thinking or the feeling." One might argue that this is *not* true of individuals' memories of observable acts – such as Ozzie and Ivy's break-in and killing in my earlier hypothetical. Far from being typically assured a high degree of privacy, the existence of such memories can often be inferred by others even without the use of neuroimaging or any other sophisticated "mind reading" technology. If I go to see the movie, *Total Recall*, with a friend, for example, I will assume (and probably assume *correctly*) that when I see my friend a few days afterwards, she will have a memory of having seen that movie, and remember its characters and plot. Moreover, unlike a private opinion or hope, our entry into the theater where the movie is playing might be recorded by an external video camera. Similarly, one might argue, when the video camera and computer chip linked to Ozzie's brain record his events in the crime, the information they capture is not information about deeply hidden internal beliefs he can be sure are unshared until he shares them: It is information that records actions that might also have been recorded by a video camera mounted in the house he was robbing. The same, one might argue, would be true of the memories that Ivy's brain makes it possible for her to create and later reexperience at a later time. Such memories of encounters with the external world may still be private to some extent – but they are not *inherently* private in the same way that unshared feeling or opinion is. The degree of privacy they should receive, one might argue, should depend not on the fact that they are memories, but on the privacy of the events someone is remembering: Police should, perhaps, have less access to an individual's memories of an intimate conversation than of someone's entry into another person's property.

Still, there are reasons we should perhaps be concerned not only what information is about – but how the government is obtaining that information. As I explain more fully later, in Chapter 5, the Fourth Amendment generally bars police from warrantlessly obtaining information from a file inside a person's home or personal computer even when that information is public knowledge and there is no constitutional bar to their obtaining it from a public source. Similarly, it may be deeply worrisome for police to learn about our past actions by forcibly "extracting" memories from our minds even if they could obtain records of the same actions from other sources using other less invasive methods.

To be sure, whether a certain use of neuroimaging violates a constitutional principle depends not only on considerations like those I have been discussing, but also on the specific constitutional doctrine at issue.

A Brief Tour of the Constitutional Privacy Landscape – and Neuroimaging's Possible Place in It

To understand better why the constitutional puzzles raised by neuroimaging are challenging puzzles, it is helpful to consider the relevant constitutional law more closely. As noted earlier, one might assume that at least in the United States, certain constitutional provisions – such as the Fourth and Fifth Amendments of the US constitution – already do stand in the way of government-compelled neuroimaging. But a closer look at these amendments raises doubts about this claim, especially when we consider them in light of the hypotheticals I have just presented.

The Fifth Amendment's Bar on Self-Incrimination

The Fifth Amendment's self-incrimination clause states that "no person...shall be compelled in any criminal case to be a witness against himself." Police may not, consistent with the Fifth Amendment, force a person facing a criminal trial to give a statement incriminating himself. And if they instead seek to obtain information about the accused's past actions from compelled brain scans rather than compelled statements, this – some scholars argue – would be an unconstitutional end-run around the Fifth Amendment (New 2008, 193–195). Even if police are not using a defendant's spoken words to incriminate him, they would be using unspoken thoughts and memories of a kind that, in prior years, would have been accessible to them only through the accused's testimony. If, as the Supreme Court has indicated, "it is contrary to the letter and spirit of the Fifth Amendment" to "force someone to disclose the contents of his mind," then forcing someone to disclose this mental content by submitting to a brain scan should be just as impermissible as forcing him to disclose his thoughts verbally (Curcio v. United States 1957, 128; United States v. Hubbell 2000, 43). Or so the argument goes.

Consider again the scenario where Ivy is suspected of committing a murder with Ozzie as her accomplice – and where the police wish to examine Ivy's brain in an fMRI scanner (perhaps while she is presented with a picture of the murder victim or the victim's living room). Ivy will argue that police's viewing of her brain's response to the stimulus is functionally equivalent to compelling her testimony. They are forbidden by the Fifth Amendment from forcing her to answer the question, do you recognize this face or this living room?, so they are instead forcing her to take a brain scan which provides police with the mental states that would underlie an honest answer to those questions. The government, by contrast, will argue that changes in the Ivy's brain triggered by her presence in (a memory of) that living room are not at all like compelled statements. They are not seeking to answer the hypothetical question of how Ivy would respond if she were compelled to answer questions about the victim or the crime scene. They are rather asking whether Ivy's brain behaves in ways that are consistent with her claim that she's never seen the victim or the victim's living room. In the sense, the brain evidence is like other physical evidence. If, for example, police found a carpet fiber from the victim's living room on one of Ivy's shoes, this would challenge her claim that she's never been in that living room Similarly, if police find her brain responds to the living room picture in a way that they know (from past experimental work) is extremely unlikely if she'd never seen it before, then this – they can argue – is a reason to believe Ivy was in that living room, and a reason that has weight no matter what Ivy would say (if forced to speak) and no matter what Ivy's current beliefs might be about what she did or didn't do in the past. All that matters is that her brain showed activity it wouldn't be likely to show unless she had a past encounter with that living room (just as it would be very hard to explain why images of the living room are on Ozzie's video recording if he'd never been there).

Whether the evidence lies in a fiber from the living room carpet, or a video recording in Ozzie's camera, or the wiring in Ivy's brain, government might argue, in all of these cases the nature of the evidence is the same for self-incrimination clause purposes: the defendants' interactions with the crime scene left a mark on their persons, and police are free to uncover such traces of defendants' pasts so long as they don't require the defendant to tell them about it. Although the Fifth Amendment establishes a zone of autonomy, in a sense, shielding the defendant from having to communicate to police, evidence of her guilt, this zone cannot be so extensive – government might argue – that it can shut the police for all evidence that the defendant's criminal activity may have left in the world, or on his own person. To do so would arguably go beyond safeguarding the suspect's autonomy, or other interests protected by the Fifth Amendment, and deprive the police of much of the evidence they, and the justice system, will respectively need to investigate and fairly adjudicate criminal cases.

As noted earlier, evidence revealing that Ivy was in a particular living room also tells police something about Ivy's memories: If evidence places her there, then it is likely she has some memory of being there. But such an inference is possible not only when police see certain brain activity in an fMRI scan, but also when they find carpet fibers on Ivy's shoes. The brain activity may provide more confidence that Ivy has a memory of the living room. But, is its linkage with the mental state that it correlates with close enough that one can justify treating the brain activity record as equivalent to a self-incriminating statement in a way that the carpet fiber is not?

The Fourth Amendment's Search and Seizure Protection

The Fourth Amendment analysis follows a similar pattern: seemingly simple on first examination, but more complicated and puzzling with a deeper look. The Fourth Amendment protects "persons, houses, papers, and effects" against "unreasonable searches and seizures." Its protection against unreasonable searches, the US Supreme Court, has said, shields Americans against "too permeating police surveillance" (United States v. Di Re 1948, 595). Although the public – and the police, on the public's behalf – has an important need to uncover criminal activity and bring its perpetrators to justice, this need must coexist with a need for privacy and spaces which, as the Court has said, remain largely free from the government's presence and control. Thus, while police are expected to investigate crimes vigorously, this does not mean they may decide, any time they like, to enter a person's house and rummage through her belongings, or log into her computer and review digital files there, in the hope they may find some evidence of a crime. Rather, in US society, unconsented-to state entry into a person's private home or files – or any other place in which she has a "reasonable expectation of privacy" (Katz v. United States 1967, 360–361) – is supposed to be an unusual event that can be justified only by an unusual circumstance; namely, a situation where police can show a judge that they have "probable cause" to believe they will find evidence of a crime in that house or those files.

The first question courts would thus have to answer in deciding whether and how the Fourth Amendment limits the scanning of an individual's brain is whether such a scan would count as a "search" subject to constitutional limits. The answer is almost certainly "yes." The Supreme Court has repeatedly held that when certain private realms – like our bodies or houses – are (as a general matter) off-limits to police surveillance, government may not circumvent such privacy protection by using "see-through" technology to observe the insides of such private realms from the outside (Kyllo v. United States 2000, 36–37). Police may not enter and explore, the interior of our homes without a warrant – and nor may they use thermal imaging devices from a street outside our homes to peer into what is happening inside (Kyllo v. United States 2000, 36–37). They may not pat down a person's coat or sift through his pockets on a whim – unless their doing so is reasonable under the circumstances (Katz v. United States 1967, 18–19). And authorities, courts have made clear, are likewise constrained by Fourth Amendment limits when they use metal detectors, X-ray devices, or similar technology to examine individuals or their coats, or the insides of their handbags or suitcases at airports, train stations, or sporting events (United States v Henry 1980, 1227, United States v. Epperson 1972, 770, United States v. Albarado 1974, 803–805, LaFave 1996, §10.7). Officials likewise engage in a search when they use X-ray technology to view the insides of our bodies.

They do something similar when they use fMRI or fNIR technology to image a person's brain activity: These devices respectively use radio waves (in a magnetic field), or near-infra red light, to gather information of processes inside a person's body that would otherwise remain hidden. This is sufficient to make such a technique count as a Fourth Amendment search.

Thus, the Fourth Amendment analysis of our prior hypothetical about Ivy and Ozzie may seem much simpler than the Fifth Amendment analysis: It would be a search for police to search Ivy's brain – even if they have no intent of inferring anything about her mental states. Any time the state gathers information from the body's interior it is a search, so state-mandated neuroimaging is a search no matter what its purposes are. And this is true even if we view the state's accessing, and monitoring of Ivy's brain activity, as analogous to obtaining and watching the video stored in Ozzie's camera and computer chip – because that video extraction would also be a search (and a seizure of the video) under the Fourth Amendment. If government helps itself to files (whether electronic documents or video files) from the inside of your computer or SmartPhone, this is a Fourth Amendment search and seizure, and so it would likewise be a search and seizure for them to help themselves to the video files stored in Ozzie's camera and computer chip.

But that the use of neuroimaging, or other advanced technology, is presumptively a search is only the beginning of the Fourth Amendment inquiry. Even if the Fourth Amendment stands in the way of government-compelled brain scans, this leaves us with a second important question – namely, just how high a barrier does it present? In other words, even if brain scan is generally such a "search" by law enforcement – and thus subject to the Fourth Amendment command that it cannot be "unreasonable" – under what circumstance may law enforcement nonetheless show that gathering data from the brain is reasonable?

As a general matter, law enforcement can only prevail in such an argument when it can obtain a warrant from a neutral magistrate, something it can in turn do only when it specifies a place, person, or thing to be searched and shows it has probable cause to believe it will uncover evidence of criminal activity there. But the warrant requirement is not necessarily a very high bar. To show probable cause, police generally need to demonstrate "a fair probability that contraband or evidence of a crime will be found in a particular place" (Illinois v. Gates 1983, 238). That provides a hurdle of sorts, but – if our unshared mental life is a sanctuary we want to insulate from government except when it has extraordinary need to enter it – we may want to demand more by way of justification than just a "fair probability" that they can find evidence of a crime there. We may want to prevent such entries except when the crime or threat law enforcement faces is an especially grave one.

Moreover, while the Constitution places warrant requirements and other procedural barriers in the way of police searches of our "persons, houses, papers, and effects," it also frequently leaves police with routes around these requirements. Police, for example, need not obtain a warrant when they use investigatory techniques where use of a warrant "would be impracticable" (like unannounced sobriety checkpoints) to meet challenges "beyond the general interest in crime control" (such as finding drunken drivers and removing them from the highways) (Vernonia Sch. Dist. 47J v. Acton 1995, 653). Consider, again, the example of airport screening at US airports: Government certainly engages in a search when it uses metal detectors, X-rays machines, or similar technology to receive information about what lies underneath the surface of our clothing or inside our bags. But it does not need a warrant to do so. Nor does it need probable cause to think that any passenger it searches has a weapon or other dangerous item. Given the security needs at stake, it is allowed to use such technology to search all travelers whether it has any reason to suspect them or not (United States v. Epperson, 770).

One might thus ask whether government could ever use the same argument to justify warrantless and suspicionless brain scanning at airports (assuming it were feasible) – or in other sites, such as transportation hubs, federal courthouses, or crowded sporting or cultural events, where law enforcement can show it needs to take robust anti-terrorism measures. Or whether it can incorporate neuroimaging into the warrantless searches police often can and do make "incident to an arrest," for example, when they require a breathalyzer test of a person arrested for drunk driving to capture evidence of his intoxication while it is still present (Birchfield v. North Dakota 2016).

As it turns out, the answers to these questions under current Fourth Amendment doctrine, generally depend on what a court finds when it balances the security (or other) interests the government is promoting against the individual's privacy interests threatened by such a search (Delaware v. Prouse 1979, 654). This is in turn, requires some sense of just how significant these privacy interests are. Is neuroimaging a significant invasion of our privacy even in its current, very limited form? Is it a threat to mental privacy if government is seeking evidence not about our unshared beliefs or feelings, but rather about what (criminal or crime-related) actions we performed in a world that we do share with others, and that police have a right to collect evidence about? To what extent, one might ask, is the nature of the privacy interest at stake in searches of the brain any different from that which is at stake in searches of a personal computer, or a cell phone or other mobile device? These aren't questions with clear answers. Until we have a better sense of how courts would analyze the privacy interests at stake in neuroimaging, we are ill-equipped to even begin to think about how they would weigh them against interests such as preventing terrorism in the air or violent attacks on schools, or catching drunk drivers or equipment operators, or preventing arrestees from harming police or destroying evidence.

And on this issue, it may well make a difference whether neuroimaging of a brain is mind-reading or brain-reading – whether it really provides government officials with information about our unshared thoughts and feelings, or instead tells them only about how a brain reacts to certain stimuli, and what this might indicate about the owner of that brain's past interactions with the outside world. It if it is the latter, a number of scholars have worried, then neuroimaging might be treated as analogous to other methods police use to gather data from individuals' bodies – such as breathalyzers or fingertip swaps – which courts have allowed police to use without warrants (and sometimes, without having to overcome any significant constitutional hurdle). It may likewise make a difference to courts that certain mental states may seem more private then others. Perhaps government is more likely to prevail if it is using neuroimaging to make inferences about a person's past interactions with the outside world than it is when using it to derive information about their past or present intentions, or sense of guilt about particular actions.

It is quite possible, of course, that judges will view neuroimaging's privacy impacts as more significant than those at stake in a breathalyzer test or fingertip swap – and thus as placing more weight on the Fourth Amendment scales, in favor of individual protection and against government freedom to investigate. If, for example, courts view Ivy's brain activity as equivalent to the video and computer chip data in Ozzie's brain – if they view the biology underlying our natural mental process as being worthy of the same privacy protection as the data in our "extended minds" – then Fourth Amendment protection may be fairly strong. Courts, after all, have recently been wary of allowing government too much leeway to easily search our cell phones or computer hard drives, so they may likewise place limits on how much government can search the natural equipment we use to create and retrieve memories.

Even here, however, Fourth Amendment protection is less certain than some would like it to be, if they wish to view mental privacy as, in most cases, invulnerable to government observation. Again, police might overcome Fourth Amendment barriers to computer or cell phone searches by obtaining a warrant based upon probable cause. They might also, in some circumstances, conduct warrantless searches of computer memory, and courts may well be willing to let them do so more readily when the search software is programmed only to turn up information connected to certain kinds of criminal activity, such as child pornography. On this model, government may also be able to neuroimage us when it can show that the mental information it seeks is likely to carry information of interest in investigating a particular crime, or thwarting a particular safety threat. In some cases, perhaps, society may need police to have access to such information in our minds. But it's also possible that in opening a door too readily for government to access this kind of needed information about our unexpressed thoughts, Fourth Amendment law may simultaneously give government access to thoughts and feelings it shouldn't be able to access.

The First Amendment, Externalized Thought, and "Freedom of Mind"

The protection the Fourth and Fifth Amendments offer for mental privacy is thus, at best, limited and uncertain. To the extent that the Fourth and Fifth Amendments may fail to protect against intensive police surveillance, it is possible the First Amendment might serve as a backstop. This, in fact, is precisely how Daniel Solove proposes courts treat the First Amendment when government engages in another, more familiar kind of "mind-reading": The discerning of our thoughts that it does by learning what books we have read, what Web sites we have searched, and the people or organizations we have chosen to associate with (Solove 2007, 112–120). As I noted in Chapter 1, these records of our intellectual life represent another kind of breach of our mental privacy. Government often does not have to peer into our brain, and make complicated inferences from the physiological measures what it detects there, to understand what is going through our minds. It can instead capture the far more easily available – and, in the present era, generally, far more detailed – records of our thinking that we leave in the outside world, about which Web sites we visit, what search terms we enter into Web site searches, or which books we download to our Kindles or iPads. Daniel Solove observes that in many situations, the Fourth and Fifth Amendments fail to protect US citizens against such government-compelled production of such records – for example, in cases where such compulsion is directed not at the citizen, but at others who hold such information, such as a bookstore or an Internet Service Provider (Solove 2007, 112). But where such government evidence gathering would "chill" speech and intellectual exploration of the kind it is prying into, courts – he says – should raise a First Amendment barrier against such efforts. Neil Richards similarly invokes First Amendment law – and its protection of freedom of thought – in arguing that constitutional law should provide robust protection for what he calls "intellectual privacy" (Richards 2015, 5). And Julie Cohen does so as well in arguing that when the state allows copyright holders to monitor our reading of their digital works, it undercuts First Amendment values and chills intellectual exploration (Cohen 1996, 1004–1019).

As in the case of the Fourth Amendment, the analogy we considered – in our hypothetical crime investigation – between Ivy's brain-based memory and Ozzie's video-based memory, cuts in favor of protection here. It would threaten Ozzie's core First Amendment interests for the government to view – whenever they wished – the video files that Ozzie creates and watches. Indeed, watching the videos of Ozzie's life would threaten his freedom of mind even more than government does when it monitors my choices about which YouTube video to watch. After all, my choice of YouTube videos informs my thinking – it contributes to it. But it is not an essential part of the way my memory or other mental process works: If my laptop and phone became damaged, I'd be very frustrated (and find life a lot harder in many respects). But I'd still have my memory, problem-solving skills and other mental capacities. By contrast, Ozzie's video is (in our hypothetical) an integral part of his memory. If the video camera or computer chip stops working, he loses much of his long-term memory capacity. So, by watching his videos, government is spying on the raw materials for his memory. And if it is a First Amendment violation for the government to spy on our use of the "extended mind," this should also be true when government spies use fMRI scans to gather information about mental processes that the brain's biology makes possible.

In fact, there are some respects in which First Amendment law (or related liberty protections in the Bill of Rights) might limit use of neuroimaging in a more straightforward way. Imagine, for example, that, as in the dystopian society George Orwell describes in 1984, some officials who hold power ask law enforcement to act as "thought police" seeking to detect and take action against those who secretly oppose the government's views (Orwell 1949, 2–4). Imagine that they wish to use neuroimaging to ferret out such hidden dissent. Or envision a variant of this government surveillance that may be more likely to occur even in a generally free society. In response to a series of anonymous posts on Twitter attacking a local official, law enforcement in that area wish to use futuristic neuroimaging (perhaps by deceptively gathering brain activity data through brain-computer interface devices) to identify (1) the individuals who have a memory of having posted these criticisms and (2) those in the town's audience who sympathize with them. These government actions could well avoid raising a Fifth Amendment problem, especially if the neuroimaging data is not obtained by government's compelling an individual to provide it, or even if compelled, is focused on individuals who would be non-party witnesses, rather than defendants, in any criminal trial. Government action could likewise avoid raising a Fourth Amendment problem (as I explain more fully in Chapter 5) if the government obtains the data it uses to infer thoughts from information that the targeted individual has voluntarily shared with a private party, or "abandoned" in Web transactions.

But such action would very likely raise a First Amendment concern – for two reasons. First, where neuroimaging is one tool that the government uses to uncover the source of speech it wishes to punish, such as the source of the Tweets in my above hypothetical, its use may encounter some of the same First Amendment barriers courts have raised against other government measures for depriving anonymous speakers of their anonymity.

As I have mentioned before, Supreme Court precedent holds that government may not compel anonymous speakers to disclose their identity – unless the government has strong interests in doing so, and cannot easily satisfy them another way. More specifically, the Court subjects such anonymity-eliminating measures to some form of First Amendment scrutiny (McIntyre v. Ohio Elec. Comm'n, 514 U.S. 334, 341–343 (1995); Watchtower Bible and Tract Society of New York, Inc. v. Village of Stratton (2002)). Lower courts have frequently found that civil litigants cannot use the discovery process to compel anonymous individuals (witnesses or defendants) to reveal their identity without satisfying certain tests designed to assure that their need for the information outweighs the expressive interests threatened by such compelled identification (Columbia v. Seescandy 1999; Dendrite Intern. V. Doe 2001; Doe v. Individuals 2008). It is less clear that such First Amendment safeguards would stop government when it learns about a speaker's identity through surveillance rather than through compelled disclosure of identity (Blitz 2005, 711–713). But given that free speech law is meant, in part, to prevent government from "chilling" speech – and to do so, even when government uses indirect means to burden speakers instead of directly censoring or silencing them, such use of neuroimaging would seem to raise First Amendment questions.

Indeed, constitutional problems of this sort might arise whenever neuroimaging is used by the government to interfere with the exercise of any constitutional right (not only freedom of speech, but also, for example, a right to keep and bear arms). If government intentionally burdens the exercise of that right by shining a light on the individual who holds it, and making her more vulnerable to government or community pressure, such a weakening of anonymity might well count as the kind of burden – on speech or other liberty – that government is barred from imposing.

And, if the intellectual privacy arguments discussed above are correct, First Amendment problems are raised not only when neuroimaging is used by government to disclose speakers' identities, but also when it is used to reveal thoughts that individuals have no intent of revealing in speech or action. The First Amendment, the Supreme Court has said, not only protects freedom of speech but also "freedom of mind." It bars government from "prescrib[ing] what shall be orthodox in politics, nationalism, religion, or other matters of opinion" (West Virginia State Bd. of Educ. v. Barnette, 637, 642). Where government seeks to reveal or expose individuals' unorthodox beliefs in order to discourage individuals from holding them, such action would seem to run afoul of this First Amendment principle. Government, the Court has said elsewhere, "cannot constitutionally premise legislation" – or presumably other government compulsion – "on the desirability of controlling a person's private thoughts" (Stanley v. Georgia 1969, 566).

Moreover, even where the First Amendment fails to offer such protection for freedom of thought, the "liberty" component of the Fourteenth Amendment's due process clause may do so. The Supreme Court has, in three cases, said that the latter constitutional provision imposes limits on the circumstances in which government can subject individuals to compelled psychiatric medication (Washington v. Harper 1990, Riggins v. Nevada 1992, Sell v. United States 2003). These three cases involved situations where authorities wished to administer anti-psychotic medications to a prisoner they deemed dangerous, or a criminal defendant in a prison or mental health facility who they wished to make competent to stand trial. Although the Court generally treated these cases as being about bodily liberty, and freedom to refuse medication, rather than about freedom of mind, it is hard to believe that the Court would find the same Fourteenth Amendment liberties entirely inapplicable if government found a method of altering individuals' minds in the same way with non-invasive techniques (instead of by forcibly administering drugs). A number of scholars have analyzed the ways in which the First and Fourteenth Amendments might ground such a "freedom of mind" and have asked about the implications a more developed constitutional freedom of mind doctrine might have for government compelled mind-alteration, or its restriction of individuals' shaping of their own mental processes (Tribe 1988, 1321–26, Winick 1989, 27–41, Boire, R.G. (Summer 2000), Boire, R.G. 2005, 234–237, 257, Blitz 2010, 1050–1057, 1069–107, Blitz 2016). In previous work, for example, I have explored the question of whether the First Amendment's "freedom of mind" protects thought only "when we put that thought into words – or some other form of First Amendment 'speech'" or does so "even when that thinking is unaccompanied by any First Amendment expression" (Blitz 2010, 1051). That work focused on cognitive enhancement or use of other methods to change thinking. But it is also conceivable, as Dov Fov argues in discussing the Fifth Amendment privilege against self-incrimination, that a "right of control over the use of their thoughts vis-a-vis the state" would be implicated not just by government interference with shaping of one's mind, but also by invasion of a person's mental privacy through compelled brain scanning (Fox 2009, 763–764, 794–797). Such a possibility clearly exists where government is invading mental privacy for the purpose of subsequently penalizing (or otherwise imposing a costs on) having unorthodox or dissenting thoughts. It may also be true in other circumstances. Thus, Adam Kolber explores whether the First Amendment protects casino users' mental privacy against laws preventing silent card counting and asks if free speech law protects the privacy of their thoughts or other thoughts (even when the though is unaccompanied by expression) (Kolber, 2016, 1382–1398). Neil Richards suggests that the need for legal protection for "intellectual privacy" flows from First Amendment freedom of thought, since the Amendment must protect not only a "marketplace of ideas," but also "the workshops where ideas are crafted," that is, the cognitive processes we use to forge them (Richards 2008, 396). Stacey Tovino likewise asks "whether a government-imposed fMRI violates an individual's privacy of thought under the First Amendment" (Tovino 2007, 460 n. 349).

These observations may provide a doctrinal answer to a question that Kiel Brennan-Marquez raises about the constitutional significance of mental privacy: Why he asks, should evidence about our private thinking be viewed as "a constitutionally special domain of evidence" – such that "a sphere of private rumination" (a phrase he borrows from Nita Farahany) should be more deserving of constitutional protection than "a sphere of private existence in one's home, or an expectation of not being arrested for no reason while walking down the street" (Brennan-Marquez 2013, 258–263). Of course, even if the value of privacy of thought were only equivalent in importance to the home or freedom from arbitrary seizure, it would still merit the Fourth Amendment protection that applies to those activities (and Brennan-Marquez doesn't deny that). But private thinking may also merit an additional layer of privacy protection, not only because it has historically been more immune to external observation or government interference than even in-home activity (as I noted in the introduction), but also because – unlike conduct with a greater distance from free expression and the formation of beliefs – it is activity that is protected by the First Amendment.

This possible application of First and Fourteenth Amendment doctrine to neuroimaging merits deeper exploration (and I intend to say more about it in future work). But I will, in the remaining chapters of this book, focus instead on another way that First, and perhaps Fourteenth, Amendment protection for intellectual privacy, and other aspects of our freedom of mind, may influence and shape legal doctrines that are more likely to be applied, in the near future, to neuroimaging in something like their current forms–namely, the Fourth and Fifth Amendment doctrines I have described above. In other words, First Amendment law may not only provide a backstop for Fourth or Fifth Amendment law that provides protection when those criminal procedure protections vanish (as Daniel Solove they may do in the searches that target writing or Internet activity rather than records of brain activity (Solove 2007, 112)). It may also shape the way those criminal procedure amendments apply to certain emerging technologies that affect our speech and thought.

In fact, the Supreme Court has taken note – in certain Fourth Amendment law cases – of the First Amendment concerns raised by a search when its target is a bookstore. In the 1965 case of Stanford v. Texas, for example, involving the search of a bookstore, it said that "the constitutional requirement that warrants must particularly describe" what authorities wish to seize must "be accorded the most scrupulous exactitude when the 'things' are books and the basis for their seizure is the ideas which they contain. No less a standard could be faithful to First Amendment freedoms" (Stanford v. Texas 1965, 480). In another case where the court found a radio show was protected by the First Amendment when it broadcast an illegally recorded conversation (on a matter of public interest), the dissenting Justices stressed that laws protecting against electronic eavesdropping not only protect privacy, but also the possibility of having personal conversations that are "frank and uninhibited, not cramped by fears of clandestine surveillance and purposeful disclosure" (Bartnicki v. Vopper 2001, 543, 553). These the dissenters noted, were "speech" interests, and that limits on electronic eavesdropping promote the "First Amendment rights of the parties to the conversation" by "protecting the privacy of individual thought and expression" (Bartnicki v. Vopper 2001, 543, 553).

These statements by judges hint at the possibility that First Amendment protections for freedom of thought may frequently work *through* Fourth Amendment protections. Rather than serving as a backstop that provides privacy when Fourth Amendment fails to – First Amendment-values might shape *how* the Fourth Amendment applies. In fact, I will argue in Chapter 5, that it already does so in certain ways: Some Fourth Amendment privacy protections (particularly in the realm of communication) seem to function a little bit more like the First Amendment protection courts give to communications than the Fourth Amendment protection they give to private spaces.

Futuristic Thinking

What makes it even harder to apply these constitutional provisions to emerging neuroimaging technology is that both sides of the intersection between constitutional law and neuro-technology are moving targets. The technology, of course, is ever-changing, and the neuroimaging of the future may well raise legal or ethical concerns that aren't concerns at all for the neuroimaging of the present. Imagine technologies of the future that allow government to make reliable inferences about our thoughts surreptitiously – by, for example, manipulating us into unintentionally revealing mental content as we interact with brain-computer interfaces. Or tools that otherwise allow government to detect, from afar and without our knowledge, that we know, and have disdain for, a particular person or are deeply familiar with a particular place. Such tools would naturally raise far more privacy worries than does the neuroimaging that takes place today. Most modern-day experiments reveal only whether a willing experiment participant in a laboratory shows brain activity – in response to a picture or phrase – that indicates some recognition of the stimulus, and may leave those conducting the tests (and perhaps law enforcement authorities they aid) with significant doubts about what this result means.

The constitutional law that scholars wish to fit to such neurotechnology is also itself always evolving. And even the stable and long-settled aspects of these legal doctrines sometimes become unsettled by new technologies. In 2012, for example, the US Supreme Court made it clear that location-tracking technology, which it had earlier held police can use in public free from any constitutional constraints, sometimes requires a warrant under the Fourth Amendment: Police may not, without a warrant, secretly attach a Global Positioning System (or GPS) tracking device to a suspect's car. Nor, said five Justices in opinions separate from the one issuing the Court's decision, can they warrantlessly use GPS or similar technology to constantly track a suspect's movements over public roadways for four weeks (United States v. Jones 2012, 954–964). Such technological changes, the Court noted, sometimes require changes in constitutional rules. In fact, according to Donald Verrilli, in an interview he gave just before stepping down as the United States Solicitor General in June 2016, modern technology – and its effect upon privacy – is perhaps the most significant engine of constitutional change: Courts have been exploring, and will have to continue exploring whether "advances in technology" require that our long-standing rules for privacy be "refashioned to meet the challenges government use of technology poses" (PBS NewsHour 2016).

What further complicates such a constitutional analysis with multiple moving parts is that such technological development is far from the only source of legal and constitutional uncertainty. Just as futurists imagine, and puzzle over, how we should deal with advanced incarnations of neuroimaging and other evolving technologies, so legal thinkers imagine – and often, vigorously advocate – different versions of constitutional doctrine. For example, they often imagine, and argue for, versions of Fourth Amendment law that protect privacy far staunchly than existing Fourth Amendment law. Or that proposes to reconcile law enforcement interests in crime fighting, with individuals' interests in security against unconstrained observation, in ways the Court hasn't endorsed yet (and perhaps hasn't yet considered). In other words, legal thinkers often imagine and propose what they deem to be "refashioning" of the kind Donald Verrilli describes before courts actually confront the need to engage it.

The challenge this book considers is thus not merely an exercise in applying current legal doctrine to a present-day snapshot of an existing technology. It is also a challenge of considering how a future, and somewhat idealized, version of First, Fourth, and Fifth Amendment law should apply to a future, and perhaps far more powerful and invasive, variant of neuroimaging technologies now emerging.

As speculative and tentative as thinking about future developments necessarily is, it is worth undertaking, for at least two reasons. One is that, while predictions about future technologies are rarely exactly right, thinking ahead can still provide valuable guidance for those who will confront them – and their constitutional implications. Fourth Amendment law has already benefitted from judges' efforts to prepare the constitutional groundwork for technologies that have not yet arrived. In 1928, for example, Justice Louis Brandeis issued a famous dissent from an opinion – Olmstead v. United States – in which the Supreme Court refused to extend Fourth Amendment limits to police wiretapping. The Fourth Amendment, Brandeis passionately argued, should not only protect individuals' privacy from the government wiretapping of his own day – it should also protect it from the surveillance technologies of the future, such as "advances in the psychic and related sciences" that "may bring means of exploring unexpressed beliefs, thoughts, and emotions" (Olmstead v. United States, 1928, 474). Brandeis's position was embraced by the Court thirty-nine years later, when it extended Fourth Amendment limits to wiretapping and electronic eavesdropping. In 2001, the Supreme Court majority also seemed to take heed of Brandeis's insistence that "in the application of a constitution, our contemplation cannot be only of what has, been but of what may be" (Olmstead v. United States, 1928, 474). It placed limits on police use of thermal imagers to measure infrared readings of a home and noted that, while the thermal imagers of the time were relatively crude, Fourth Amendment law should not "leave the homeowner at the mercy of advancing technology." The thermal imaging technology or other wall-penetrating technology of the future, it noted, might "discern all human activity in the home" (Kyllo v. United States, 2000, 35–36).

Second, futuristic thinking in the law not only prepares courts to deal with future facts – it also often helps courts and legal scholars to frame important questions about present-day constitutional law. Consider two examples of this in Fourth Amendment scholarship. In 1984, Arnold Loewy explained how a fictional "divining rod" could help us understand the Fourth Amendment's underlying purposes: He asked his reader to imagine that there was technology that allowed police to equip themselves with a perfectly accurate evidence-detecting instrument a police officer "would walk up and down the streets and whenever the divining rod detected evidence of crime, it would locate the evidence. First, it would single out the house, then it would point to the room, then the drawer, and finally the evidence itself" (Loewy 1983, 1244). Society would thus be spared erroneous searches which not only waste the time of the police, but also subject innocent people to harrowing and disruptive government intrusions into their homes and files. Loewy was aware that such a divining rod was a thing of science fiction. But it provided a useful ideal, he argued, for how the Fourth Amendment should try to make police searches work, one which might guide courts as they dealt with less fanciful surveillance technologies. It should, he argued, limit the rummaging police do through the parts of our lives that deal with innocent activity and do not reveal the crimes police are investigating (Loewy, 1983, 1247–1249).

Indeed, as explained more fully in Chapter 5, the same year Loewy published this idea, the Supreme Court found what it took to be an example of a real-life divining rod in trained police dogs that alert only to the presence of illegal drugs and to nothing else. A canine sniff of a personal package, said the Court, does not violate any legitimate privacy interest, because it detects only illegal drugs which no one has a right to possess or hide (United States v. Place 1983, 707).

In 1996, a law review note by Michael Adler then imagined another version of a Fourth Amendment "diving rod" – this one for the age of cyberspace: It envisioned a scenario where law enforcement had the capacity to send software throughout the Internet that could identify digital contraband – such as a child pornography video or stolen intellectual property (with copyright protection stripped off of it) – with perfect accuracy on anyone's computer (Adler, 1996, 1097–1098). Individual computer users would be completely unaware of such a digital search while it was occurring, and – so long as their computers were found to be free of such contraband – would be left undisturbed by the government. The search, Adler posited, would have "a minimal impact on property, produc[e] no false positives, need not be noticeable, and revea[l] nothing to officials beyond the identity of some individuals who possess this particular piece of digital contraband" (Adler 1996, 1100). Thinking about how the Fourth Amendment applies to such technologies is valuable not only because of what it tells us about how the U.S. Constitution might apply to future technologies that are functionally similar – but also because of what it tells us about the Fourth Amendment and its fundamental purposes. Does it exist only to prevent police intrusions into the homes or property of innocent people? Or does it protect private spaces – like the computer drives in this hypothetical – even when they contain illegal content?

Asking similar constitutional questions about brain imaging devices – whether about fMRI, fNIR, or EEG devices akin to those that exist now, or about powerful variants of these devices that may or may not arise in the future – helps us to explore whether and how the U.S. Constitution should protect the privacy of our unexpressed thoughts and feelings. To the extent the Fourth Amendment protects the privacy of these thoughts and feelings from government brain imaging, we can ask, does it do so only accidentally – as a side effect of protecting our brains and the rest of our bodies? Or should it also shield our thoughts in order to better secure a space where our intellectual and emotional life is under our own control – and insulated against government control or monitoring – not only when we generate thoughts in our brains, but also when we generate them with technology outside of our brains, whether with a pen and paper, or an Internet search engine? To the extent it shields, from government observation, information about our mental processes, just what does this cover? Does it prevent the police from drawing inferences about our conscious perceptions and other thoughts – and then sharing them with the jury? Or does it also shield unconscious mental processes? Does it shield information about enduring characteristics of our mental lives (such as whether we are typically shy or extroverted), or only discrete memories, beliefs, and feelings?

CHAPTER 3

Lie Detection, Mind Reading, and Brain Reading

Abstract This chapter briefly looks at the ways that those in the mid-to-late twentieth-century developed lie-detection techniques without neuroimaging – and how various neuroimaging techniques promise more sophisticated types of lie detection. It also very briefly explains how different neuroimaging technologies – such as EEF, fMRI, and fNIR – work, and how they might evolve into more sophisticated – and invasive – techniques in the future, and how law enforcement use of them may thus raise privacy concerns (and do so, even in cases that at first seem free of substantial privacy harms).

Keywords Electroencepholography · EEG · Functional magnetic resonance imaging · fMRI · fNIR · Guilty knowledge · Lie detection

LIE DETECTION BEFORE NEUROIMAGING: POLYGRAPH TESTS

People have long understood that that certain physiological phenomena provide a visible component of certain emotional experiences (such as feeling fear, shock, or surprise). Anxiety, for example, may manifest itself in increased sweating or shaking. Shock may result in pale appearance, and embarrassment may cause blushing. As Francis Shen writes, "humans...are natural mind readers," who have long used various strategies, such as "using facial

© The Author(s) 2017
M.J. Blitz, *Searching Minds by Scanning Brains*,
Palgrave Studies in Law, Neuroscience, and Human Behavior,
DOI 10.1007/978-3-319-50004-1_3

expressions and body language to gauge intent" (Shen 2013, 658). In fact, systematic approaches to ferreting out dishonesty have existed far longer than neuroimaging. As Sarah Stoller and Paul Root Wolpe write, "[t]he development of a successful lie detector has been a dream of governments and law enforcement since ancient times" (Stoller & Wolpe 2007, 359).

The use of scientific machinery for lie detection is more recent. In 1921, John Augustus Larsen, a police officer and medical student, invented the polygraph – a machine that uses physiological readings, typically of blood pressure, heart rate, respiratory rate, and electro-dermal activity (electric activity due to skin's secretion of sweat) (Alder 2007, 6–9). Since that time, moreover, some police and private organizations have sought to improve this biology-based mind reading with technology: They have used polygraph machines to measure changes in blood pressure, galvanic skin response or muscular activity in the hopes such changes could reveal when subjects are lying. In recent decades, some scientists have sought to perfect such physiology-based lie detection. Paul Ekman, for example, has proposed that people display involuntary facial "micro expressions" – "very brief facial expressions, lasting only a fraction of a second" – when they "deliberately or unconsciously concea[l] emotions," and that these can provide clues (but not proof) that a person may be speaking dishonestly, and trying to conceal their nervousness or discomfort in doing so. (Ekman, Micro Expression, at http://www.paulekman.com/micro-expressions/).

These familiar methods of lie detection already provide one model for a kind of mind reading: An official can make an inference about someone's unexpressed mental state – and, particularly, whether she believes what she is stating, or whether she finds a particular word or image more significant than others – by measuring that person's heartbeat, blood pressure, and other physiological indicators. In fact, when Ronald Allen and M. Kristin Mace asked readers to imagine a mind-reading example – in order to explore if police use of it would violate the self-incrimination clause – the hypothetical they present describes lie-detection methods modeled on those familiar from polygraph machines – one where, even as a suspect refuses to cooperate, and indeed, remains stubbornly, silent, the machine operators "record[s] ... changes in his heart rate, blood pressure, breathing, and electro-dermal responses (electrical conductance at the skin level)," that occur as he is presented with specific information about a crime (Allen & Mace 2004, 248–249).

To be sure, these methods of mind reading by lie detection have been met with great skepticism. Some observers argue that the most common lie detection methods don't provide anything close to an accurate measure of honesty. Moreover, the nervousness many people experience in such a test situation may well lead to false positives in spite of the controls, and countermeasures can be taken by which a guilty respondent may be able to control his emotions (Hughes 2014). As Daniel Langleben and Jane Campbell Moriarty note, "most U.S. courts have expressed disapproval of polygraph-based evidence...and courts remain largely hostile to its admission into evidence" (Langleben & Moriarty 2013, 223).

Still, it is worth looking more closely at the two most common variants of lie detector tests before we look more closely at neuroimaging techniques – because many uses of neuroimaging as lie detection technology have adapted these methods. Lie detection using a polygraph has typically followed one of two paradigms. In the "control question test," or "comparison question test," individuals are interviewed and then asked a series of questions, each of which falls into one of three types. The questions to which law enforcement typically needs an answer, in the criminal context, are "relevant questions," which are questions about what occurred in a crime (such as, did you kill the victim?). But, if a person subjected to such a question manifests the physiological signs of nervousness or other uncomfortable reactions one would expect to see in someone who would rather not answer, it may be not because she actually committed the crime, but rather because simply being asked a question (in the context of a criminal investigation) is unsettling, and perhaps triggers the fear she is under threat from an authority even if she did nothing criminal. So, the test also asks "comparison" or "control" questions, questions which are also designed to feel threatening, although they are not about the crime. Such questions frequently explore other behavior individuals would be reluctant to admit – for example, whether they have misled friends or spouses, or acted unethically in school or work situations. The premise of asking these two types of question is that the physiological responses for each question will be distinct – so researchers can distinguish questions where individuals merely feel threatened (the control questions) from those where they are lying about their innocence in a crime (the relevant questions). Finally, individuals are also sometimes asked irrelevant questions – questions about mundane issues (Is your name Marc? Are you wearing a white shirt?) that don't threaten the subject and in which the physiological reaction should be at the baseline level (Committee to Review the Scientific Evidence on the Polygraph, National Research Council 2003, 14–15).

While the control question test has been the most widely used method, it has been criticized for uncovering the subject's anxiety when answering in response to questions rather than any reliable proof of whether the subject participated in a crime. David Lykken wrote in 1959, for example, that such lie detection relied on "unreasonable assumptions about the consistency of physiological response patterns" (Lykken 1959). A better alternative, he argued, is to test not for dishonesty or deception, but for "guilty knowledge." As a consequence, some researchers advocate a different kind of method of detecting when someone has participated in criminal or other wrongful activity – the "Guilty Knowledge test" or "Concealed Information Test."

The Guilty Knowledge Test (GKT) presents a subject with multiple choice questions, the answer to which only a participant in a crime could know – for example, a multiple choice question that asks what the murder weapon was, or how criminals in a home invasion and armed robbery made their way into the home, or what kind of vehicle they used as the get-away car. Only someone guilty of the crime should be able to identify the correct answer ("the probe") among the various options, since, as one study notes, the "neutral" or "control" alternatives are carefully chosen so that an innocent subject – lacking knowledge of the crime – would not think they are any less likely to have been involved than the "probe." If an individual consistently shows a greater physiological response when presented with the "probe" than when presented with the controls, it is likely because he has knowledge of the crime (Committee to Review the Scientific Evidence on the Polygraph, National Research Council, The Polygraph and Lie Detection 2003, 15). A 2003 study noted that this kind of test has been quite successful in discriminating between those with guilty knowledge and those without it – although it also noted that "almost all attempts to examine the validity of GKT were based on simulations (i.e., mock crime experiments) in which some participants (the guilty) are required to commit a mock crime (e.g., to steal an envelope containing a sum of money and piece of jewelry from a specified office)" (Carmel et al. 2003, 261–262).

Some studies have argued that the guilty knowledge method of lie detection is far superior to the comparison control question method, but critics have still worried that it is susceptible to countermeasures – and also lacks accuracy even when countermeasures aren't used. The use of neuroimaging for lie detection – especially EEG and fMRI – thus strikes some as a possible solution, particularly if such neuroimaging technology improves to the point of overcoming certain challenges that exist in transferring these methods from laboratory experiments to law enforcement use.

LIE DETECTION WITH NEUROIMAGING: METHODS OF MONITORING THE BRAIN IN ACTION

Many other books and articles have provided explanations of how brain-based mind reading techniques have worked in the laboratory. I will not discuss them with the same level of detail here. But a little background on the issue is useful for understanding how the law might make use of them – and what constitutional implications they might have (which I will look at more closely in the subsequent chapters).

Each of the 100 billion or so neurons in a human brain participates in a massive and complicated exchange of signals by repeatedly generating an "action potential," that is an electrico-chemical change that generally begins in branch-like structures called "dendrites," which receive chemicals released by other neurons. The action potential travels over the membrane of the cell, down a wire-like filament called an "axon," at the end of which it causes the release of chemicals that travel across a "synaptic gap," separating the neurons – and, by doing so, either increasing or decreasing the likelihood that another neuron down the chain will "fire." Somehow, the complex ensemble of action potentials that neurons generate and send to each other makes possible our conscious awareness – and the many aspects of mental life than go with it: Our capacity to have feelings, to remember past events, or imagine what the future would be like (or conjure images and sounds that make up purely fictional scenarios).

While no one knows why or how this neuronal action produces conscious awareness, scientists don't have to know this to find a pattern in brain activity that seem to consistently arise together with a certain type of mental experience. The question raised for neuroimaging-based lie detection, then, is this: Can neuroimaging identify patterns in brain activity that consistently arise when someone is dishonest, possesses "guilty knowledge," or has some other mental state that those employing lie detectors wish to uncover? If so, can it do so more reliably than more traditional methods of lie detection (including the methods human beings use when they judge another person's honesty without the aid of technology) – and reliably *enough* for police or courts to admit them into evidence?

Such questions, however, will not necessarily have the same answer for all methods of neuroimaging. One method of neuroimaging that has been promoted as a basis for lie detection is electroencephalography (EEG). It measures the brain's electrical activity via a set of electrodes placed on the surface of the scalp. Rather than measuring the electrical activity of individual neurons (which is done only with the invasive insertion of tiny microelectrodes – essentially small glass pipettes – beneath the skull), EEG measures the electrical currents that arise in regions of the brain as many millions or billions of neuron fire in sync – in coordinated rhythmic generation of action potentials. This rhythmic activity produces "brain waves" the character of which varies with different kinds of activity. Each electrode detects the electrical signals from the underlying region of the brain, and sends the signal to a machine that amplifies it.

EEGs have been used for decades. The German psychiatrist Hans Berger invented the technique in 1929, while seeking a scientific explanation for how telepathy might transmit ideas from one person's brain to another (Buszaki 2006, 5). While he failed to produce evidence for the possibility of such natural mind reading, his invention has, since the late 1980s, been tested as a possible means of artificial mind reading of a kind – and more specifically, as a substitute for older lie detectors (Langleben & Moriarty 2013, 224).

The most widely discussed form of EEG lie detection has recreated a variant of the "guilty knowledge" or "concealed information" test which, instead of using electrodermal or other traditional measures, uses a kind of brain wave pattern called a "P300 wave." Scientists have found – in many different types of experiments – that the P300 pattern appears when these subjects view specific kinds of stimuli but not others. Electrical activity generated in the parts of the brain in response to a specific stimulus is called an "event related potential." Scientists often label certain parts of the wave form generated in such a potential with a "P" if the wave's amplitude is positive and an "N" where the wave's amplitude is negative – followed by the number of millisecond seconds that elapse between the time the subject is shown the stimulus and the time that such a high or low point in the wave appears (Ward 2010). A P300 wave is thus a positive peak that occurs in many individuals roughly 300 milliseconds (and generally within a range of 250 to 400 milliseconds) after a perception of a certain kind of stimulus (Ward 2010; Sur & Sinha 2009 Jan–Jun). Scientists have found that P300 is often elicited by an "oddball" stimulus – that is a stimulus that a subjected is instructed to watch for that occurs relatively infrequently and differs markedly from a more frequently presented stimulus (Wolpow & Wolpow 2012, 216–127).

Researchers have thus hoped that among the stimuli that will trigger a distinctive kind of P300 is an item or scene that strikes them as familiar (and, if it is unlikely to be known by those other than a perpetrator of a crime, would constitute "guilty knowledge"). J. P. Rosenfeld, who initiated the study of EEG-based lie detection in the 1980s, has recently conducted experiments testing the usefulness of P300 measurement to reveal hidden knowledge related to a mock crime. In 2010, he and John Meixner had subjects envision themselves being a part of a terrorism group and write a letter to a fictional terrorist leader about the method, timing, and location of the bombing. When later presented with words from the letter, the subjects showed stronger P300 signals than when reading other words about times, places, and possible weapons (Hughes 2014). More recently, Rosenfeld and Meixner performed another study, which elicited strong P300 waves when they showed subjects words related to activities these subjects had recorded the previous day (while wearing a video camera that recorded for four hours as they followed their normal routine) (Hughes 2014).

The most well-known (and heavily-promoted) use of EEG as a lie-detection method is "brain fingerprinting" – which Lawrence Farwell developed using his own variant of P300-wave measurement: Farwell measures what he calls a "P300-MERMER" signal. This variation uses not only the amplitude peak but also a negative peak that closely follows the positive peak (Farwell 2012). Like the other EEG variants of the guilty knowledge test, brain fingerprinting, as Farwell describes it, is a method of detecting "[w]hen an individual recognizes something as significant" in the "context" of the testing. When an individual does have such recognition, the "'Aha!'" response he has in this circumstance is accompanied by the P300-MERMER response (Farwell 2012). Farwell has promoted his technology as ready for courts, but other researchers have voiced skepticism, given the proprietary nature of the technology (Brandom 2015).

Since 2000, scientists have also explored whether they might use functional Magnetic Resonance Imaging (fMRI) to detect lies, other dishonest behavior, or concealed information of interest to the justice system. fMRI does not measure the electrical signals of neurons directly. Rather, as Jones, Buckholtz, Schall, and Marois state, "[i]n much the same way that the body delivers more oxygen to muscles that are working harder, the body delivers more oxygen to brain regions that work harder" (Jones et al. 2009). It does so by delivering more oxygen-rich blood to neurons that are firing more actively than they usually do. fMRI uses a combination of magnetic fields and radio wave bursts to map this flow of oxygenated blood within the brain. In short, it magnetically aligns the spin of hydrogen atom protons throughout the brain, knocks their spin out of sync with radio waves, and then measures the energy they emit as the magnetic field brings the atoms back into alignment: This energy emission is different depending on the oxygen level of the blood, so scientists can use this difference to create a "blood oxygen level dependent" or BOLD signal that varies with the level of oxygen. It cannot create such a measurement for each neuron: Rather, it divides the brain into "voxels" – a three-dimensional equivalent of a "pixel" – each of which consists of about a million neurons (Yuhas 2012).

In recent years, fMRI has received far more attention than EEG methods because of the startling information scientists have been able to infer or reconstruct on a screen – such as pictures and words a person was imagining, or episodes they were remembering or dreaming. As Langleben and Moriarty write, those who think about lie detection have also begun to focus more heavily on fMRI because "recent progress in the ability of fMRI to reliably measure and localize the activity of the central nervous system has created the expectation that an fMRI-based system would be superior to both the polygraph and the EEG for lie detection" (Langleben & Moriarty 2013, 223). fMRI is not superior to EEG in all respects, however: As they note, while fMRI "is greatly superior to EEG in its ability to localize the source of the signal in the brain," EEG "is significantly less expensive, more mobile, and has a better time resolution than fMRI," as it is measuring electrical activity in the brain as it changes and not relying on blood flow as a proxy (Langleben & Moriarty 2013, 223). In coming years it is possible that at least some of the EEG's advantages – such as its portability – can be combined with some of fMRI's ability to localize brain activity: Functional near infrared imaging (fNIR) makes use of the same premise as fMRI – that is, it measures neuronal activation by measuring the flow of oxygenated blood within the brain. But instead of using a combination of powerful magnetic field and radio waves – requiring a room of heavy equipment – to trace the flow of oxygenated blood, it does so by sending specific wavelengths of "near infrared" light into an individual's cortex and measuring how the light is absorbed by the brain tissue (Ayaz et al. 2011). This neuroimaging technology is in some respects more limited than fMRI (it can only detect changes up to 4 cm deep in the brain tissue), but is far more portable and able to measure individuals as they engage in routine activities: measurements require that individuals wear a set of probes and headgear, something they can do as they move around (Ayaz et al. 2011).

In any event, certain fMRI devices – like EEG tests – have had some success in revealing deception or "guilty knowledge" in experimental settings. Neuroscientists have worked since the early 2000s to test fMRI's success on this front in various scenarios. These studies generally showed that certain activity would increase in certain brain regions when a subject was engaged in deceptive behavior. In 2001, Sean Spence and his colleagues conducted a study in which subjects were instructed to tell the truth or lie, while under a scanner, in response to certain questions about their lives (Spence 2004). When lying, the subject's brains showed greater activation in certain areas: the ventrolateral prefrontal cortex and the medial prefrontal cortex. In 2002, another experiment conducted by Lee found that certain brain areas showed greater activation when individuals feigned poor memory (Lee et al. 2002). And Langleben conducted another fMRI lie detection experiment in which subjects – who had been given playing cards – were instructed to deny only their possession of one of those playing cards (one which they had received with $20 in an envelope) when asked if they had particular cards. Thus, they would dishonestly deny having that particular card, but acknowledge having the others they were given. This experiment too found that certain areas of the brain (in this case, the anterior cingulate cortex) showed greater activation during deceptive behavior (Langleben et al. 2002) (and this result may stem not only from the mental acts' deceptive behavior but also from the distinctive reaction to the card the subject recognizes as the one accompanying the envelope).

As Sarah Stoller and Paul Root Wolpe write, more recent neurotechnological lie detection has moved from focusing on brain regions to "the study of general patterns of brain activation distributed over many regions of the brain," a technique that "could enable more accurate predictions of cognitive and affective states" (Stoller & Wolpe 2007, 361). Others have done fMRI experiments that closely parallel the kind of guilty knowledge or concealed information tests that, in EEGs elicit a P300 response. For example, one study published in 2012 found that "probes were consistently accompanied by a larger percentage signal change than irrelevant items" in a test asking them to view playing cards, one of which they had selected in an envelope (Gamer et al. 2012, 509).

These lie detection tests, of course, raise significant questions about reliability – and about how jurors will react to presentation of fMRI, EEG or other neuroimaging evidence. As Langleben and Moriarty note, some of the models have been critiqued on the grounds that they measure not dishonesty of the kind one would find in the real world but rather behavior (and accompanying cognition) that occurs when "subjects are explicitly instructed to lie" (Langleben & Moriarty 2013, 224). There have also been questions about whether fMRI lie detection that works in situations where the stakes are low (for example, where students can keep money if they lie successfully) will work in real life situations where someone faces conviction and imprisonment.

As I noted in the Chapter 2, these are among a number of concerns that scientists and other scholars have raised about use of neuroimaging to reveal deception or "guilty knowledge." Some also raise questions about the use of group data to make an inference about an individual. Barbara Sahakian and Julia Gotwald note that before fMRI can be "used in courts for lie detection, we will need larger studies that determine the levels of accuracy of the technique, especially at the individual level" (Sahakian and Gottwald 2017). They note that experimental methods that work on one set of subjects, may not work on people with different characteristics, and that aspects of real-world life detection may make techniques that successfully uncover instructed deception in laboratory settings ineffective in other settings. They also take note of a 2011 study by Giorgio Ganis, in which brain scans were able to tell, with 100% accuracy, when an experiment subject was lying about not recognizing the date of their birthday – but also found that these subjects could "disrupt the model" in the experiment, and thus make it difficult to tell false from true answers, by taking simple countermeasures, such as moving a finger or toe (Sahakian and Gottwald 2017).

Other neuroimaging technologies might likewise be used to make inferences about honesty, or about other mental states and activities. Like fMRIs, Positron Emission Tomography (PET) and Single Photon Emission Computer Tomography (SPECT) scanning can measure blood flow within the brain and thus determine what regions become more active in particular tasks. PET and SPECT scanning do so by injecting radioactive isotopes of a molecule (like oxygen) into a test subject's blood, and then detecting gamma rays produced from the collision of positrons (from the radioactive isotope) with electrons nearby. (SPECT and PET differ in the radioactive isotopes used as "tracers".) As Francis Shen notes, these methods "have [ ] been used in a variety of criminal and civil cases" (Shen 2016, 501).

Another neuroimaging technology – magnetoencephalography (MEG) – has better temporal resolution than fMRI because, like EEG, it measures neurons' electrical activity not indirectly, by measuring blood flow, but rather by measuring the magnetic fields generated by this electrical activity (MEG's spatial resolution, however, is poorer than that of fMRI) (Snead 2007, 1282, Sahakian and Gottwald, 2017).

NEUROIMAGING BEYOND LIE DETECTION: SCREENING MOVIES FROM THE MIND

My focus here is not on all of the legal questions that might be raised about this technology (such as whether they can produce admissible evidence), but rather on the possible privacy implications of these technologies. On the one hand, use of an fMRI to test for deception – or to see if someone recognizes an object presented by a test administrator – seems far less threatening to privacy than the mind-reading devices of science fiction, that can pull memories out of individuals' minds. As Fox notes, it makes a constitutional difference that such technologies are "not capable of exposing the content of a subject's cognitive thoughts" – and instead give government access only to "the less privileged sphere of sensory recall and perceptual recognition about a particular set of facts or the state of past events" (Fox 2008, 2). Moreover, it is the research study itself that will typically supply the content initially: In the comparison question test and any neuroimaging analogues of it, it is the researcher who asks the questions – with the subject giving a simple "yes" or "no" response.

Still, it is conceivable that such technology can – even in something close to its present state – be used in ways that threaten individual privacy. First, even a single yes-no question, or multiple choice question designed to elicit guilty knowledge, might undercut privacy when it concerns an issue that a person regards as highly sensitive (such as that person's sexual behavior, or whether they have a certain medical condition). Second, if officials have the opportunity to pose *many* yes/no or multiple choice questions – and learn something from a subject's EEG or fMRI response even where the subject refuses to answer – they might learn a good deal about the subject's private life. Third, neuroimaging may well be used in conjunction with other investigatory tools, which themselves present a threat to privacy. For example, an investigator may not need an fMRI or EEG test to tell where a subject has been, or what she has done in the past week, if she has access to a week's worth of location-tracking information, video footage, or text messages and e-mails. In a case like that, fMRI and EEG might simply fill in some blanks – for example, about person's attitude toward people and places in the records authorities already possess. In fact, even where an official has significant doubts about an inference she has drawn from neuroimaging evidence, she might conceivably use such evidence as a starting point (or intermediate step) in an investigation in which she later confirms an educated guess, based on neuroimaging, with evidence from other sources.

Still, future development in neuroimaging technology can substantially change the nature of the privacy threat it poses. Lie detector-style neuroimaging does not, of course, provide individuals with the detailed information about memories or thought that one finds in a personal diary. Such detail would require mind-reading technology akin to that in science fiction where investigators turn brain activity that correlates with internal thoughts into verbal descriptions, videos, or some other representation of a person's memories, beliefs, or dreams.

This is not something that current neuroimaging technology can do. But there have been small experimental steps in that direction. These generally work by using many fMRI readings of individual test subjects to create a dictionary of sorts, each entry of which matches specific fMRI readings with a particular act of cognition or perception. Researchers, for example, might establish what pattern of brain activity (as seen in an fMRI machine) arises when one is thinking about a particular word, such as "house," or viewing a particular image, or a specific person's face, or hearing a voice with certain characteristics. Then, when they see a previously categorized fMRI reading (or something very like it), then can infer that the person is likely to be experiencing something similar to the matching perception or thought.

The process is, of course, a complex one – and relies not only on the power of fMRI machines, but also on advances in computer technology, as it is computer algorithms that match the complex fMRI readings with specific cognitive tasks.

One research team, headed by Jack Gallant, was able to use fMRI technology to reconstruct simple images – and then, in a later experiment, videos – that someone was watching from fMRI readings of brain states. In the image experiment, a computer was able to guess, with a high level of accuracy, which image (in its library of images) a person was likely viewing. The video translation worked on the same principle – but went further: The research did not simply "guess" what parts of a video clip someone was watching (based on a match with fMRI readings taken during previous viewings of the video), it also used a computational model to predict what brain activity patterns would arise as individuals watched other movies (not yet viewed by the subject) – and then used this model to reconstruct additional videos individuals were watching, as they watched them, from the sequence of BOLD signals it detected in the fMRI readings (Nishomoto et al. 2010; Smith 2013). Another study was similarly able to use fMRI readings of activity in the brain's hippocampus to tell where in a virtual-reality environment an experiment subject was, as they navigated through it (Chadwick et al. 2010). Still another study allowed fMRI machines to reconstruct the faces people were viewing. This could conceivably allow witnesses to reconstruct the face of a potential culprit by lying in an fMRI machine and imagining a face instead of describing it to a sketch artist or picking it out of a photographic line-up (Cowen et al. 2014).

These experiments reconstructed images from brain activity that occurred as the individual was viewing an image or watching a video. But researchers have also been able to guess or reconstruct what people are imagining in their "mind's eye." Marcel Just and Tom Mitchell were able to use fMRI technology to identify what patterns particular words and concepts generate in a person's brain activity when someone prompts them with the word or concept, and to develop models that can predict what kind of fMRI pattern would arise for other words (Shinkareva et al. 2008). Another group of researchers using fMRI instruments was able to determine which of several film clips someone had seen. And a group of researchers in Japan was even able to produce "dream recordings," using fMRI readings to determine – with 60% accuracy – what objects people had reported seeing during dreams. In fact, they were able to reconstruct crude videos of the dream's imagery based on these measurements.

Science fiction has already envisioned how far more advanced variants of this technology might work in a futuristic legal system. In the movie, *Strange Days* (1995), for example, a key witness to a murder is herself killed, but a recording of her memory remains available (and persuasive, in a world where the technology for creating such recordings is well-established). Obviously, if and when neuroimaging should ever become capable of extracting from our mind such vivid movies of our past experience, or transcripts of our silent thinking, the privacy threat it presents will be far graver than it is with technology of the kind that exists now.

It is also possible that instead of revealing what people are thinking or remembering at a particular time, neuroscience technology will instead reveal certain enduring features of people's personalities. Other biological research – for example, on DNA features – may already provide scientists with methods for making inferences of this kind. For instance, some studies have correlated shyness and anxiety with having a particular variant of the "serotonin transporter promoter" gene (Battaglia et al. 2005, 85, 91). Thus, for example, while character traits are normally inadmissible under the Federal Rules of Evidence in US courts, where character evidence *is* admissible, it is conceivable one could offer evidence of this kind not only through lay testimony, but also through biological evidence of character (where courts allows experts to present scientifically-informed evidence of character traits).

Neuroimaging evidence may also be a source of such information. Various studies have correlated differences in personality with differences in a person's "connectome." Just as the "genome" is the entire sequence of nucleotides in your DNA," a connectome, says Seung, is "the totality of connections between the neurons in a nervous system" (Seung 2012, preface). To the extent this pattern of connections partly makes us who we are, should fMRIs be able to discern information about it, they might be able to uncover features of our personal predispositions. Certain studies have used a variant of fMRI called "resting state fMRI" – which looks at how the brain acts when we are resting as opposed to performing a specific cognitive task – to find correlations between certain personality types with certain brain patterns (Adelstein et al. 2011, 4–5). fMRI evidence revealing information about subjects' responses to particular stimuli might also allow inferences about their personalites: An article by Martha Farah and her colleagues, looking at whether fMRI studies might raise privacy concerns, notes that "functional neuroimaging is, indeed, already capable of delivering a modest amount of information about personality, intelligence, and other socially relevant psychological traits" (Farah et al. 2010, 126). It also notes that it is conceivable that scans that an individual takes for one reason (such as measuring face perception) may also incidentally collect information about other aspects of the subject's psychology (for example, because "extraversion and unconscious racial attitudes are both correlated with brain activity evoked by simply viewing pictures of faces") (Farah et al. 2010, 110).

CHAPTER 4

The Fifth Amendment: Self-Incrimination and the Brain

Abstract This Fifth Amendment's self-incrimination clause has been at the center of constitutional discussions over neuroimaging's future. That it is not because it clearly would apply to neuroimaging – but rather because neuroimaging raises a easily formulated (albeit difficult to answer) Fifth Amendment puzzle: It seems to count as both of what are supposed to be two mutually exclusive categories in Fifth Amendment law, because it is both like a witness statement (or "testimonial") and like physical evidence such as blood flow or other physiological processes. This chapter explores various solutions scholars have proposed to this puzzle, rooted in distinctive theories of the self-incrimination clause – and the unanswered questions each of these theories raises. It also emphasizes another point that has received less attention in discussions of self-incrimination and neuroimaging: idea that Fifth Amendment protection for our thoughts and other mental process should perhaps sometimes cover the biology underlying that thinking even when government plausibly claims it wants access to it for reasons other than inferring our thoughts or beliefs.

Keywords Fifth Amendment · Mind reading · Neuroimaging · Self-incrimination · Testimonial · Witness

© The Author(s) 2017   59
M.J. Blitz, *Searching Minds by Scanning Brains*,
Palgrave Studies in Law, Neuroscience, and Human Behavior,
DOI 10.1007/978-3-319-50004-1_4

An Overview: Why Neuroimaging Raises Fifth Amendment Problems

The Fifth Amendment questions raised by neuroimaging's possible law-enforcement applications have received more attention from scholars over the past decade than other constitutional questions, and this is not surprising. It is in Fifth Amendment self-incrimination law that neuroimaging most clearly raises a constitutional categorization problem of a kind deeply familiar to courts: Is mental content "pulled" from a criminal defendant's mind with the aid of brain-based mind-reading like a verbal statement or other testimony that government may not compel the defendant to provide? Or is it more like the physical evidence – like blood samples or fingerprints – that may be compelled by government?

In a 1966 case called Schmerber v. California, the Court made clear that this question was central for self-incrimination law. It stressed that "the privilege is a bar against compelling 'communications' or 'testimony'" (Schmerber v. California 1966, 764). When the government does not compel the defendant to say or communicate anything, but instead treats him as "a source of 'real or physical evidence'" – such as a fingerprint, or a blood test – then, said the Court, it does not violate the defendant's Fifth Amendment privilege against self-incrimination, even if the physical evidence government obtains from the defendant is just as damaging for the defendant's prospects in a trial as would be a confession or other self-incriminating statement (Schmerber v. California 1966, 764). So just where does neuroimaging fit in this testimonial-physical dichotomy?

One reason it is clear that this is a question that may well be asked about neuroimaging is that the US Supreme Court has already come very close to asking it – in Schmerber itself. Today's neuroimaging technology did not, of course, exist 50 years ago when Schmerber was decided. But polygraph lie detection devices were already in use, and the Supreme Court digressed from its discussion of blood testing to consider where they might fall in the testimonial-physical evidence dichotomy:

> There will be many cases in which such a distinction [between testimonial and physical evidence] is not readily drawn. Some tests seemingly directed to obtain "physical evidence," for example, lie detector tests measuring changes in body function during interrogation, may actually be directed to eliciting responses which are essentially testimonial. To compel a person to

submit to testing in which an effort will be made to determine his guilt or innocence on the basis of physiological responses, whether willed or not, is to evoke the spirit and history of the Fifth Amendment. (764).

While this statement begins by portraying physiological testing in lie detection as a kind of gray area, it ends by strongly suggesting that "the physiological responses" evoked by such methods would count as testimonial. The privilege against self-incrimination, it immediately added (quoting a prior US Supreme Court case), must be "as broad as the mischief against which it seeks to guard." And the clear import of the Court's words here would be that coaxing physiological responses from a person who refuses to provide verbal responses, would be an example of this "mischief" – a kind of trickery whereby the government gets from a person's body the answers to questions it is constitutionally forbidden to obtain from his compelled communication.

THE PURPOSES OF THE SELF-INCRIMINATION CLAUSE

But later scholarship on the Fifth Amendment implications of neuroimaging has not simply accepted the Court's determination. In the first place, it has raised questions about precisely what mischief the Fifth Amendment self-incrimination clause seeks to guard against – and the purposes it is designed to promote. According to many scholars, those purposes are entirely unclear. Akhil Amar and Renee B. Lettow observe that the clause has always "lacked an easily identifiable rationale" and has been "unsolved riddle of vast proportions, a Gordian knot in the middle of our Bill of Rights" (Amar & Lettow 1995, 857). William Stuntz likewise writes that "most people familiar with the doctrine surrounding the privilege against self-incrimination believe that it cannot be squared with any rational theory" (Stuntz 1988, 1228). And Ronald Allen and M. Kristin Mace similarly note that "the theoretical foundations of the Fifth Amendment are conventionally thought to be in disarray" and that many of explanations of its purpose often given by the Supreme Court are problematic. The Court, for example, has suggested that the amendment shields a criminal defendant in order to protect the "inviolability of the human personality and of the right of each individual to a private enclave where he may lead a private life" (Murphy v. Waterfront Comm'n 1964, 55). But as Allen and Mace point out, "law molds and shapes 'human personality' directly, constantly and unavoidably; and immunity permits the most private aspects of a person's life to be divulged, as occurs in criminal and civil cases daily across the land" (Allen & Mace 2004, 244). The state can compel witnesses other than the defendant to testify in criminal trials, and can even compel the defendant himself to testify if it grants him immunity – so if privacy is the central value of the Fifth Amendment's self-incrimination clause, it is one Fifth Amendment doctrine does not seem to protect effectively. It continues to allow the state to "demand evidence from every area of our personal lives" (Allen & Mace 2004, 262).

Another explanation the Court has offered is that, without the shield of the self-incrimination clause, an individual could be subjected to what the Supreme Court calls a "cruel trilemma" – wherein, when forced to answer the government's question under oath, he is forced to choose between "self-accusation, perjury or contempt" (Murphy v. Waterfront Comm'n 1964, 55). But this trilemma does not confront innocent defendants (who could answer questions honestly without self-accusation). As Stephen Schulhofer notes, "the innocent defendant faces no trilemma, no dilemma, in fact, no problem at all" (Schulhofer 1991, 318). If one faces a trilemma of this kind, it is, as Stuntz points out "the consequence of" having committed the crime one is questioned about (Stuntz 1988, 1239).

It is problematic then to rely on the purpose of the self-incrimination clause to tell us where – in the dichotomy between physical and testimonial – we should place a seemingly border-line evidence-gathering method: if we can't identify the clause's purposes, we lack such a categorization device. We could conceivably proceed more modestly, and proceed by analogy. Even without identifying the underlying purposes of the self-incrimination clause, for exam- ple, we might be able to show that neuroimaging evidence is sufficiently similar to clear-cut cases of physical evidence that compelling it from a defendant belongs on the same (permissible) side of the doctrinal line. Thus, Sean Thompson looks at prior case law and argues that based on such law, "evidence obtained with fMRI scanning is physical evidence" (Thompson 2007, 347). Or we might, by contrast, analogize the brain activity in response to a stimulus to a testimonial statement: To the extent an observed event (or set of events) in our brain occurs in response to an interrogator's question, and is treated by that interrogator as functionally-equivalent to a verbal answer, the perhaps it should be classified as testimonial – regardless of the underlying purposes of the self-incrimination clause (Holloway 2008, 166–174). Brennan-Marquez challenges such an analogy, arguing that it is not how the interrogator treats the responses that is essential, but rather the intent of the defendant: The defendant has to have an intent "to convey information, above and beyond being stimulated in a way that simply produces information" (Brennan-Marquez 2012–13, 253). In any event, analogy with clear-cut cases provides one way of proceeding.

Another is to try to find some underlying self-incrimination purpose that can guide us – and scholars have tried to do so in one of two ways. Some have attempted to supply a logic for the self-incrimination clause that succeeds in making sense of it where the Court's accounts have failed. William Stuntz, for example, offers an account of Fifth Amendment purposes different from those mentioned above, an account based on excuse in criminal law: Faced with the choice of whether to lie (and commit perjury) or confess to a crime that carries significant punishment, most individuals would feel significant pressure to lie, and would likely give into it – so a guilty criminal defendant should not be penalized for feeling, and giving in, to the same pressure.

The law should excuse him in the way that it excuses other unlawful choices when the pressures to make them are such that normal individuals could not easily resist (Stuntz 1988, 1248–1260). Stuntz offers his theory as a more normatively attractive, and more coherent, explanation of the self-incrimination clause than those offered by the court.

Michael Pardo proposes another principle as a lodestar for the self-incrimination clause law: the principle that "government may not compel for use as evidence, the content of a suspect's propositional attitudes" (Pardo 2006, 330). When government does so, he argues in another article, it forces the defendant to serve as an "epistemic authority" for the fact-finding that will determine his guilt: To justify their conclusions about whether the defendant was or was not at the scene of a murder, for example, and did or did not pull a trigger, juries will be relying, at least in part, on what they deem to be the defendant's actual beliefs about the answers to these questions (Pardo 2008, 1035). But forcing a defendant to share such beliefs (or try to lie about them) to a jury is, says Pardo, at odds with the presumption of innocence in criminal trials. Under the presumption of innocence, it is the government's responsibility – not that of the defendant – to "fil[l] up the epistemic void" that exists prior to the introduction of evidence. Government must make its case finding the defendant guilty by offering its own source of knowledge to the jury, rather than by forcing the defendant to offer his own knowledge of what occurred as a basis for such a guilty verdict. (Pardo 2008, 1035).

Other scholars try to derive self-incrimination clause purposes not from first principles, but rather by constructing a story about those purposes which (however appealing or unappealing it might be in its own right), at least has the virtue of being consistent with, and perhaps providing a coherent principled explanation for, the self-incrimination clause case law. Instead of trying to provide an attractive normative account of the self-incrimination clause's underlying values, these accounts focus on "positive theory" that explains what the court has done. Indeed, accounts like the two I just discussed – the excuse theory and a theory that bars government from taking and using a defendant's cognitive content – have found support not only as attractive normative accounts of what self-incrimination law should be, but also as positive accounts of what it is.

Thus, in her analysis of neuroimaging and self-incrimination, Nita Farahany begins with the excuse theory, not because it provides what she regards as the most attractive account of how self-incrimination clause doctrine should apply to neuroimaging but rather because "the excuse-based model provides the best positive account of how self-incrimination cases are decided." (Farahany 2012a, Incriminating Thoughts, 366). Allen and Mace likewise argue that while there is no general theoretical justification for the Fifth Amendment, "there is a powerfully explanatory positive theory" (Allen & Mace 2004, 245–246). But they believe that a different model of self-incrimination clause law fits best with the decided cases – and it is one that is closer to Pardo's account than to Stuntz's: In short they argue, the rule that is consistent with most self-incrimination cases is that "government may not compel disclosure of the incriminating substantive results of cognition that themselves (the substantive results) are the product of state action." (Allen & Mace, 247).

APPLYING EXCUSE-BASED THEORY OF SELF-INCRIMINATION

Let us take a closer look, then, at how each of these theories fares in helping explain the status of neuroimaging data in self-incrimination clause law. First, Nita Farahany applies Stuntz's excuse theory to neuroimaging. Doing so, she says, is useful not because it helps us to place such evidence within the testimonial-physical dichotomy, but because it should lead us to abandon this dichotomy as a guiding framework for determining what compelled evidence is incriminating. We should replace that dichotomy, she suggests, with a new set of categories for dividing up the "spectrum" of neuroscience evidence. (Farahany 2012a, Incriminating Thoughts, 366). A better classification system, she says, consists of the following four categories. First, some neurological evidence is "identifying evidence." It consists of information that is "static and descriptive," much like a person's "physical likeness," height, weight, DNA or blood type. It allows investigators to link their suspect to a place or a previous reported observation. (Farahany 2012a, Incriminating Thoughts, 368–369). For example, if they found a defendant's blood or DNA in a location, this could indicate he was there, or if a witness reported seeing a person who had a beard and was approximately 5′8″ tall, they could match the defendant's appearance to that description. A structural feature of the brain, or an enduring characteristic of its operation (like a repeated brain wave pattern) might count as identifying.

Second, other neurological evidence is "automatic evidence." This, says Farahany, consists of "evidence produced automatically rather than through conscious thought processes." She includes in this category, autonomic body functions such as "[b]linking, the beating of a heart, sweating," and also automatic "visceral or emotional reactions to external events" (Farahany 2012a, Incriminating Thoughts, 373). Moreover, she notes, there are various possible methods by which law-enforcement investigators might use neuroimaging or other technologies to detect "unconscious perception of emotionally-salient stimuli" (Farahany 2012a, Incriminating Thoughts, 375).

The third category of evidence is "memorialized" evidence: Memories, are, of course, often stored in written form – in diaries or notebooks, for example. But "places and things in one's autobiographical history have neural representations," and – as noted earlier, law-enforcement officials, of course, have great incentives to obtain such memories from a defendant if they can do so. They may want to find a way to see if, contrary to earlier claims, he knew another person they have in custody in a murder case. Or whether they might even somehow extract memories of the crime itself (Farahany 2012a, Incriminating Thoughts, 383–384).

Farahany's fourth category of evidence is "utterances," by which she means not only verbal statements, but also other actions in which a subject consciously responds in some way – even if with silent mental activity – to a question. She describes a number of neuroimaging techniques that would evoke "utterances" even when the subject says nothing: Neuroimaging techniques that cause subjects to recall certain events (and perhaps reveal evidence in brain activity that scientists can use to decode and "screen their memories"); tests that reveal other words or images that arise in the subject's mind, or that reveal subjects' internal mental states after each question in a series of questions demanding a "yes/no" answer (Farahany 2012a, Incriminating Thoughts, 389–401).

Because Farahany weaves her self-incrimination clause analysis around Stuntz's excuse-based account, she determines the Fifth Amendment status of each of these forms of neurological evidence by asking if it is the kind of evidence that, like traditional compelled testimony, a criminal defendant would have an overwhelming (and excusable) tendency to choose to falsify or distort in order to avoid punishment. On this basis, she finds that her first three forms of neurological evidence – identifying, automatic, and memorialized – aren't covered by the privilege. Defendants have little chance to disguise identifying characteristics (they can't easily falsify their DNA) and, when they have an incentive to do so (for example, by disguising physical appearance), they will "be tempted to do so absent any government compulsion" (Farahany, 2012a, Incriminating Thoughts, 371). Because automatic neurological evidence occurs automatically and outside a defendant's consciousness, here too, normal people will lack the control they need to present a false front for government observers (Farahany 2012a, Incriminating Thoughts, 378–379).

Memorialized evidence, by contrast, is evidence that individuals have often created consciously, but by the time the memory is stored and created, the individual's control has already been exercised. This kind of evidence could conceivably be obtained through neuroimaging. In the future, it may be possible for certain neuroimaging technology to retrieve the memory in a way that by-passes an individual's control – and thus, any chance to consciously reshape it. Stored memories, in other words, are analogous to memories recorded in a diary or computer file that law-enforcement agents recover in defendant's house after they have arrested him and brought him to the station: As much as he might wish he had avoided writing certain statements in such a notebook, when these memorialized writings are out of his hands, he cannot change them. Similarly, he may be unable to simply will the deletion or falsification of his episodic memories, and the government might be able to retrieve them (Farahany 2012a, Incriminating Thoughts, 381–382).

By contrast, "utterances" are – as their label suggests – clearly testimonial. But this is not simply because they are in some sense analogous to conscious communications. After all, as Brennan-Marquez points out, some conscious mental acts of a defendant may carry no communicative intent at all (Kiel 2012–13, 253–254). It is rather because even in silent evoked utterances, individuals may well have a powerful incentive to try to hide their memories or distort their thoughts. To be sure, a defendant may find it extraordinarily difficult to suppress a thought in the face of mind-reading technology – more difficult than to stay silent in response to a question. But he may, like the hero in the movie, *Village of the Damned* (1960), find a way to will himself to generate different memories or thoughts than those he knows his interrogators are looking to accuse him with – and, so long as these memories and thoughts occur consciously, he may feel strongly tempted to try. In short, even when he remains silent, neuroimaging could put the defendant in a position where he tries to "suppress his memory, create a false one, or accurately recall and potentially self-incriminate" (Farahany 2012a, Incriminating Thoughts, 401).

Here, he has an opportunity he does not have when government seeks to access his memories in ways that by-pass his conscious control: Where the government elicits "active memory recall," rather than unconscious generation of memories, the defendant is a knowing participant in the process of eliciting a memory and may try to influence how it occurs. (This assumes, of course, that the excuse model would continue to apply to situations where such temptation for falsification occurs even though falsifying unspoken thoughts, unlike false speech under oath, does not make one guilty of perjury).

In other words, rather than focusing on whether evidence is physical or testimonial, Farahany's focus is instead on whether the evidence a defendant is being asked to produce is (1) evidence he can control, and conceivably hide or falsify, and (2) if so, whether it is the prospect of being required to produce it for government that will create the almost irresistible pressure to engage in such evasion or distortion. Farahany does not claim we should be satisfied with leaving identifying, automatic, and memorialized evidence in our minds free for the government to take. On the contrary, she says, a "future where unconscious emotions, dispositions, and memories can be detected without running afoul of the privilege against self-incrimination is an alarming one" (Farahany 2012a, Incriminating Thoughts, 404). But with alternative privacy-based accounts of self-incrimination unlikely to replace an excuse-based model, she argues, the Fifth Amendment is a poor bet for providing protection for the "cognitive liberty" that the excuse-based model leaves vulnerable. The better bet, she argues, is a certain conception of Fourth Amendment law (which also differs from the dominant doctrinal model, but has a chance of gaining more judicial endorsement) and, better yet, a system of statutory protection like that which Congress has enacted to protect genetic information (Farahany 2012a, Incriminating Thoughts, 406).

One reason she is pessimistic about Fifth Amendment protection for any category other than utterances is that existing case law seems to support such pessimism. Courts haven't decided neuroimaging cases yet.

But in other situations, the Supreme Court has made it clear that identifying evidence is non-testimonial – and is open for government to force from defendant. In United States v. Wade, decided one year after Schmerber, in 1967, the Court found it was permissible for the state to make everyone in an identification line-up (including the defendant) reenact part of the crime. In the crime itself, a bank robber with a small strip of tape on each side of his face had pointed a gun at the teller and said something like "put the money in the bag" (United States v. Wade 1967, 220). So, each person in the line-up, including the defendant, was asked to wear strips of tape on their face and repeat those words. These compelled words, the Court decided, were not "testimonial": The Defendant was required "to use his voice as an identifying physical characteristic, not to speak his guilt" (United States v. Wade 1967, 222–223). In other words, the witness would predictably use characteristics as the pitch and timbre of the defendant's voice, and perhaps other aspects of his manner of speaking, to match him to the perpetrator. It was not the content of the words itself that provided any evidence – after all, they weren't really the defendant's words at all, but rather a script provided by law enforcement to him and others in the line- up. For similar reasons, the Court found, in Gilbert v. California – a case decided the same day as Wade – that a handwriting sample is non-testimonial. "[L]ike the voice or body itself," it is "an identifying physical characteristic" (Gilbert v. California, 1967, 267). The government is interested in how the defendant writes (and more specifically, whether he writes like the perpetrator does) and not in *what* he writes. Given such cases, it would not be surprising if courts similarly decided that neuroscience evidence is admissible when it reveals evidence of how someone thinks – and whether it matches what is known about the thinking patterns of the perpetrator – rather than *what* they are thinking.

Likewise, the case law on "memorialized" evidence seems to support Farahany's sense that it would be left unprotected by the Fifth Amendment: Government might constitutionally use the content of a criminal defendant's speech when that speech was created voluntarily by the defendant, rather than on the state's orders. For example, before his arrest, a criminal defendant might enter records related to the crime in a tax report or computer file. If he then leaves these writings with others, government can later subpoena them – and although it is, through such a subpoena – obtaining records of a defendant's own words, this does not mean doing so violates the Fifth Amendment. The self-incrimination clause, the Court said in Fisher v. United States, cannot "serve as a general protector of privacy," and the state may obtain "private information" where it obtains it "not...through compelled self-incriminating testimony," but from "from other sources" (Fisher v. United States 1976, 401). In fact, as the Court has made clear, a person himself "may be required to produce specific documents even though they contain incriminating assertions of fact or belief because the creation of those documents was not 'compelled.'" To be sure, added the Court, such a required act of production may itself "have a compelled testimonial aspect." Citing the Court's prior case law, the Hubbell opinion noted that in producing documents demanded by government subpoena, a witness would "admit that the papers existed, were in his possession or control, and were authentic" (Unite States v. Hubbell 2000, 35–36) Where government does not already know the answer to these questions (in other words, that they are a "foregone conclusion" (Fisher v. United States 1976, 411)) – where it needs to rely on defendant's admissions that are implicit in his act of production – then the act of production itself may be a testimonial statement covered by the privilege. In fact, the Court held the defendant's act of producing documents was testimonial in this way in Hubbell itself. But where government makes use only of the papers themselves, and not in beliefs of defendant implicitly confirmed by an act of production, it is acting permissibly under the Fifth Amendment.

Perhaps then, Farahany suggests, government could likewise get other stored information it can somehow retrieve (through inference) from the defendant's brain, so long as it can do so without forcing the defendant to consciously produce it himself. Of course, where an individual is compelled by the state to conjure up mental content that the state can't obtain on its own – then this is unconstitutional self-incrimination as it would be when an individual is forced to produce documents in a way that reveals information. But in that case, the latter case, we are in the realm of utterances and not memorialized information.

AN APPROACH BASED ON COGNITIVE CONTENT – AND SOME QUESTIONS ABOUT ITS SCOPE

Some of those who reject the excuse theory come to different conclusions – and, as a general matter, are not so pessimistic about the extent to which existing self-incrimination doctrine can cover neuroimaging evidence or other similar evidence that can be used to infer thoughts. Allen and Mace, for example, argue that compelled submission to mind reading would violate the Fifth Amendment, and that this would be true any time government is essentially requiring a defendant to provide the "substantive results of cognition" (Allen & Mace 2004, 247). Pardo and Patterson (drawing in large part on Pardo's earlier-described account of self-incrimination doctrine) similarly argue that the self-incrimination clause protects a defendant against being forced to provide any information about the "content" of his "mental states" for government to use against him (Pardo & Patterson 2013, 165).

Thus, when a defendant is forced to provide the government with the content of his memories of participation in a crime – so that the government can argue to a jury that the memories are reflective of reality – this is a violation of the self-incrimination privilege, they say, regardless of whether the defendant does so through active memory recall, or by being made to somehow produce them in a way that by-passes his conscious control. For example, they say, were a witness forced to undergo hypnosis – and then reveal memories to a jury under such hypnosis – this would eliminate any temptation she might feel to distort or hide these memories as she revealed them. Under Farahany's excuse-based analysis, they note, the self-incrimination privilege would not apply: The witness will not be tempted to hide or distort her memory of the truth, because – while under hypnosis – she doesn't even realize she is revealing it, let alone understand how doing so is incriminating herself. But, in Pardo and Patterson's view, this is still a blatant violation of the self-incrimination clause. In fact, they argue, it is a reductio ad absurdum showing that the excuse-based analysis cannot be right: Eliminating a criminal defendant's consciousness of what she is saying about a crime cannot plausibly leave the state free to make her say it, and incriminate her with what she (unconsciously) says (Pardo & Patterson 2013, 174). In Pardo and Patterson's view, "evidence generated by [compelled neuroscientific] tests would be testimonial whenever its relevance depends on the content of defendant's mental states, in particular, the content of her propositional attitudes" (Pardo and Patterson, 2013, 167). And this can cover examples of neuroimaging evidence Farahany categorizes as "memorialized" (and perhaps even "automatic") and not just those that she classifies as "utterances."

This model of self-incrimination seems to offer far more constitutional protection against compelled use of neuroimaging evidence than does Farahany's elaboration of the excuse-based model. Unlike Farahany's account, it protects "automatic" and "memorialized" evidence whenever it is used by government to convince the jury that the defendant has a certain mental content that supports a finding of guilt (such as a memory of having been at a crime scene). While broader in that sense than Farahany's account, it is, in one sense, narrower: Whereas she finds the privilege against self-incrimination covers all "utterances" (whether they occur in speech or silent contemplation), Pardo and Patterson exclude from the privilege's coverage any utterance that is used to prove something other than defendant's mental content. They illustrate this point with the case of Pennsylvania v. Muniz, in which the Supreme Court considered a challenge to prosecution's use of a defendant's statement in a drunk driving trial. The Supreme Court plurality in Muniz, they argue, was mistaken in classifying as "testimonial" and thus barred by the Fifth Amendment – the defendant's statement, responding to a police query, that he did not know the date of his sixth birthday. That statement by the defendant was, of course, an utterance (Pennsylvania v. Muniz, 1990, 586, 592–600). But it was not an utterance used to incriminate him (since, while it may be unusual to be ignorant of one's birthday date, there is nothing illegal about such ignorance). Rather, his inability to identify his birthday was used to show not that the defendant had a memory, belief, or piece of knowledge supporting a finding of his guilt, but rather that he was in a certain mental condition (that he was drunk) (Pardo and Patterson, 2013, 167).

Pardo and Patterson further illustrate how their account would apply to neuroimaging with a comparison between two hypothetical uses of neuroimaging evidence, one concerning a defendant named "Winston," who is being prosecuted for bank robbery, and another concerning a defendant named "Alex," who is being prosecuted for criminal fraud. Here is one of the two comparisons they offer of how these defendants might be subjected to compelled neuroimaging in ways that trigger self-incrimination clause protections for one defendant but not for the other:

"Winston, [] suspected of bank robbery, is [] compelled to sit for the 'brain fingerprinting' test. He is shown images of the bank vault (which only employees and the robber have seen) and presented with details of the crime. The government wants to introduce the test results, which suggest prior knowledge when presented with the images and details, as evidence of Winston's guilt."

"Alex, [] suspected of fraud, claims that he has a short-term memory problem, which explains his conduct, rather than an intent to commit fraud. The government compels Alex to sit for the 'brain fingerprinting' test. They first present him with some details, and after a short period of time, test him to see if the results suggest 'knowledge' or 'memory' when he is presented with the details. The government wants to offer the results as evidence of guilt, arguing they show that Alex did recognize the details in the test and does not have the memory problems he claims" (Pardo and Patterson, 2013, 167–168).

After these two tests, they explain, "Winston would be able to invoke the privilege" because "the 'brain fingerprinting' evidence" is relevant only in so far as it reveal the content of his mind: "[I]t provides evidence of Winston's knowledge of the crime scene and details of the crime." By contrast, Alex could not invoke the privilege because the brain fingerprinting is being used to show Alex has a mental capacity (for short-term memory he claims to lack) and not to show that his memory has any particular incriminating content (Pardo and Patterson, 2013, 167–168).

This different treatment may at first seem odd because – in both cases – the brain fingerprinting is undermining a defendant's case that they lack a certain kind of memory capacity. Winston claims (or would likely claim) that he lacks the capacity to form a memory of the bank vault because he never saw it. Alex claims that he lacks certain short-term memory capacity more generally. However, there is an important difference for self-incrimination clause analysis, under Pardo and Patterson's account because, in showing Winston has capacity to form a memory of the bank vault, the government will be asking the jury to find that Winston's memory of seeing that vault is a memory they can rely upon in reaching a guilt verdict (and thus, presumably, asking it to treat Winston as an epistemic authority).

A closer look at this account of the self-incrimination clause, however, reveals that it may be narrower than it first appears – and that it would likely fail to cover a use of neuroimaging evidence very similar to that in the Winston example. Recall the hypothetical argument by police or other government officials I imagined in Chapter 2, that they might conceivably use neuroimaging not to draw inferences about mental states, but rather about brain activity that would be unlikely to occur but for a particular past interaction with the world. More specifically, they might argue that – when their EEG device reveals a P300 wave in the electrical activity within Winston's brain (when Winston views images of the bank fault) – the EEG is generating a finding that it would be far less likely to generate if Winston had never before seen the bank vault.

In other words, government officials would be foregoing (on their own view) any reliance on mind-reading. The defendant's mental states, they might argue, are entirely inaccessible to them: They cannot know what the defendant thinks or believes, since they do not have any kind of direct access to his consciousness. Nor do they know what he would say about what he thinks or believes if he could be forced to testify (which he cannot be). What they *can* know, they might argue, is when individuals who have seen a certain object, or heard a certain voice, or are confronted with an image of that object or voice, their brains tend to show a certain activation pattern. What law enforcement is doing, they claim, is not presenting evidence of any kind regarding what the witness thinks or believes, but rather what happened – and the imprint this occurrence left in the defendant's brain (not mind).

The defendant is hence not an "epistemic authority," as it is not the defendant's knowledge or beliefs that are being offered to the fact-finders as proof of what occurred. It is the physical evidence about brain activity that occurs in response to a certain stimulus (Indeed the defendant might, if he testified, fervently argue that what he knows and believes is at odds with the neuroimaging evidence, and should trump it).

We might refine this argument by recalling the variation of it set forth in Chapter 2 with the help of the "extended mind" theory. As I noted earlier, Chalmers and Clark argue that the physical correlates of mental activity may sometimes lie outside the brain. To prove their point, they ask the reader to imagine a person, Otto, who, because of damage the early stages of Alzheimer's have caused to his memory, uses a notebook to preserve – for his mental use – information he would otherwise quickly forget (like the address of the Museum of Modern Art), and consults the notebook when he needs to retrieve the information to visit it with his friend, Inge (Clark & Chalmers 2008, 226–227). For Otto, the notebook serves a function identical to the mental function that his friend, Inge, performs when she retrieves the same information by triggering certain brain operations. Otto's notebook, they argue, therefore serves a function parallel to Inge's biologically enabled memory (Clark & Chalmers 2008, 226–227).

As I noted in Chapter 2, we might imagine a variation of the Otto and Inge hypothetical where two similar individuals – Ozzie and Ivy – rob a mansion and murder its resident. Like Otto, Ozzie has a long-term memory problem. To correct for this problem, he uses a tiny video camera implanted in his skull, which peers out and constantly records events in the outside world – and then transmits the footage to a small computer chip implanted in the brain. Ivy, by contrast, needs no such neural prosthetics: she relies entirely on brain and has an excellent long-term memory even for fine details.

A video camera, of course, operates very differently than does natural memory, as does a computer chip. But so does a notebook. On Clark and Chalmers' model, if Ozzie uses the video camera in a way that is functionally equivalent to the way that other people draw upon natural brain processes to generate memory of events – if, for example, he invariably consults it when he wonders what happened on a recent day – then the video camera's digital memory is, in a sense, a part of machinery that allows Ozzie to generate and retrieve memory. But if police then arrest Ozzie for the robbery and murder at the mansion, and make a digital copy of the video footage in Ozzie's video and computer chip, they could probably do so without thereby violating the self-incrimination clause. Although the video is an integral part of the combined technological-biological apparatus Ozzie uses to recall his memories, it would still – intuitively – probably count as physical rather than testimonial evidence. Jurors could view the video without assuming that it is merely a proxy for what Ozzie would have said on the witness stand. They can themselves form a judgment about its accuracy (perhaps with the help of an expert witness).

And so it makes sense to ask whether any of the biological activity that normally underlies memory – such as the biological activity in Ivy's brain, or for that matter the biological activity revealed by the brain fingerprinting techniques in Pardo and Patterson's Winston hypothetical – could similarly count as physical evidence for the same reason. Is there any evidence that neuroimaging could draw from her brain's operations that would be the equivalent of the digital footage in Otto's camera? To be sure, the biology of human memory is starkly different from the technology in a video camera. Against the common perception that memory is "like a videocamera," Joyce Lacy and Craig Stark note that natural memory has "substantial malleability" (Lacy & Stark 2013, 649–650). Amanda Pustilnik notes that "memory is a process not a thing," and so it may be misleading to view memories as lying in storage, waiting to be played (2013, 6). It thus may be extremely difficult to draw memory from a subject unless one asks him (or somehow presses him) to produce it by generating certain thoughts. In that case, it may be unconvincing for the government to argue that the brain evidence from such a requested or elicited mental state is evidence that can be disentangled from cognitive content, and presented to the jury without giving them any evidence of the latter. Pardo and Patterson observe that "[m]emories are not like photographs or pictures that exist in the brain and that can be extracted like files in a computer or filing cabinet. The content and production of memories cannot be so easily separated'" (Pardo and Patterson, 2013, 173). That we can imagine a cyborg-suspect of the future who uses photographs or computer files in his memory formation process does not mean that there is anything analogous in evidence drawn from the biology of actual human memories, that prosecutors may obtain without compulsion or introduce for jurors (and experts) to evaluate, without any assumptions about the defendant's reliability as a source of knowledge.

But nor is it clear that it is therefore impossible for government to by-pass the witness's knowledge or beliefs. Consider Farahany's description of how fMRI may, in the future, be used to connect a suspect to another culprit, or connect him to past events at a particular crime scene: "fMRI imaging of the auditory cortical activation patterns of a listener," she notes, "could allow investigators to identify a speaker to whom the listener has previously been exposed, or the content of a sound to which he has been exposed" (Farahany, 2012a, 381). It is at least possible to imagine, and understand, an argument in which prosecutors introduce such evidence and tell jurors it supports a finding that the defendant heard a certain voice, or heard certain words – regardless of whether he consciously remembers the experience. It may be that such an argument is unlikely to succeed, given limits on what the technology might reveal about the brain. It may also be that something crucial would seem to be missing in the prosecutors' story if they tried to tell such a story without at least implying that the fMRI evidence is revealing defendant's past actions by revealing his memories and knowledge. But such questions do not rule out the possibility that neuroimaging might count as physical evidence, under the self-incrimination clause, of our past interactions with the world and the way those interactions have affects our brain's operations.

One possible response to this is to concede that, in any plausible view of the self-incrimination clause's purposes, there will be certain kinds of detailed neuroimaging information about our encounters with the world that government will be permitted to obtain from the defendant and introduce, as long as it presents this evidence for the right kind of conclusion (that is conclusions about physical interactions in which the defendant has been involved, and not the subjective experiences he had while such interactions occurred, or those he has now).

Another is to resist any argument by the government that such a use of neural evidence would really skirt arguments about mental states. Thus, Dov Fox seems to suggest that, since "most sophisticated operations of mind are deeply integrated with the mechanical operations of biological organisms," it is implausible for the government to argue (or for individuals to feel) that evidence that probes deeply into the latter could steer clear of the former (Fox 2009, 795). Consider again an fMRI scan that, when Ivy is shown a picture of the crime scene, displays activity correlated with past presence in the scene shown in the picture. One might argue that any conclusion that she was actually *at* the crime scene still must implicitly rely on an assumption that her unspoken memory of presence at the scene is an accurate one – and would inevitably be using Ivy as an epistemic authority for the conclusion she was actually there.

A BIOLOGICAL BUFFER ZONE FOR MENTAL PRIVACY

Before ending this chapter's discussion of the self-incrimination clause, it is helpful to consider another possible response to the argument that certain uses of neuroimaging should count as permissible "brain reading" evidence rather than impermissible introduction of mental state evidence. Sarah Stoller and Paul Root Wolpe argue that it may be implausible to expect that individuals in modern Western society can so easily separate mental events from brain activities that happen in tandem with them – and to leave government with free evidentiary access to the latter so long as it leaves the former alone. "The connection that we feel to our brain," they observe, "is unlike the connection that we feel to any other aspect of ourselves. Even if the firing of our neurons is just a chemical reaction in response to a stimulus as is our bleeding in response to the touch of a needle, we still feel a more intimate connection to the activity in our brains than to the activity in our blood vessels" (Stoller & Wolpe 2007, 371). While thought, belief, or memory is not equivalent to the brain activity that accompanies it, it is still the case – they argue – that an intrusion into the brain feels to many people like an intrusion into the "actual 'self.'" Quoting neuroscientist Donald Kennedy, they note that an intrusion into the brain, may bring an outsider "way too close to who I am" (Stoller & Wolpe 2007, 372).

If this is right, perhaps the Fifth Amendment self-incrimination clause should prevent the government from obtaining certain evidence from our brains that could potentially be used to draw – and advance – conclusions about our mental states, regardless of the conclusions the government actually intends to draw from them. In other words, the Constitution might not only generate a constitutional force-field of sorts to protect against brain-based mind reading – it might also create a buffer zone of sorts that generally keeps government at an adequate distance away from the brain activity individuals now closely identify with their "actual 'sel[ves]'."

One approach that already advocates something like this comes from Dov Fox. Echoing Stoller and Wolpe's link between the brain and "actual self," Fox notes that "our thoughts are what anchor each of us as an individual person with an uninterrupted autobiographical narrative" (Fox 2009, 796). Because of this, "mental control" is an essential interest – and one the Constitution has to vigilantly guard. Moreover, Fox argues against a constitutional framework that would allow the government to use brain-imaging evidence by claiming that it is steering clear of mental phenomena and focusing only on the brain's physical responses to the world. As noted earlier, he argues the integration of brain and mind is such that it is implausible for government to argue (or for individuals to feel) that evidence that probes deeply into these operations could avoid harming our mental life as it does so (Fox 2009, 795). However, while Fox finds this right to mental control in a right to silence, it is broader than that. It is also an important dimension of the First Amendment's protection of freedom of thought, and as such, I will argue, also important in Fourth Amendment cases that deal with searches of our thinking, and use of neuroimaging to search our brains. (Indeed, Fox argues that the testimonial-physical distinction itself "presupposes a flawed conception of mind/body dualism") (Fox, 2009, 765, 795, 801).

One might well respond to this proposal by noting that it seems to treat the Fifth Incrimination Clause's purpose as that of protecting individual autonomy or privacy, a purpose (as noted earlier) that scholars have found to be a poor fit because the Fifth Amendment seems to protect such autonomy or privacy so narrowly. But other accounts of the self-incrimination clause purpose's may also be an imperfect fit with case law: It is not only the defendant, but also other witnesses, who may feel powerful pressure to alter testimony (to help a friend or relative), for example, and even if the state is blocked from introducing evidence of a defendant's knowledge or beliefs through compelled testimony or neuroimaging, it might do so by introducing written records or diary entries.

The fact that an account of the Fifth Amendment purposes is "over-inclusive," that is, that it would justify protection not just for the acts that the Court has actually protected, but also for those not covered under Fifth Amendment law, is not a decisive strike against it. As William Stuntz points out, in his article, Privacy's Problem and the Law of Criminal Procedure, some of the considerations that undermine treating privacy as an underlying value in the self-incrimination context similarly undermine the claim that the Fourth Amendment is about privacy protection. As he writes, if we view Fourth and Fifth Amendment law as about privacy protection, this raises the puzzles of why we don't see the same privacy protection in other interactions between citizen and state: "A privacy value robust enough to restrain the police should also prevent a great deal of government activity that we take for granted – activity that, at least since the New Deal, is unquestionably constitutional" (Stuntz, 1995, 1017). That the court forges ahead and protects privacy vis-a-vis police in the Fourth Amendment context suggests that privacy can be staunchly protected in some constitutionally demarcated spheres of state-citizen interaction even when it is not protected in others where one would expect to be equally justified.

Nor is this suggestion of reviving a role for privacy, and more generally, for autonomy – in elaborating the contours of the self-incrimination law's potential application to neuroimaging – necessarily at odds with the Amendment's treatment of papers, records, and other externalized thought. For one thing, one could argue that some of these instances of externalized thought ought to be protected. Amar and Lettow observe (in their own account of Fifth Amendment "first principles") that a defendant's diary testifies: "[I]t speaks in the defendant's own words, much as would the defendant himself as a witness on the stand" (Amar and Lettow 1995, 883). Even if such externalized, and fixed, thought remains unprotected by the self-incrimination clause, perhaps a different rule should apply our internal thinking processes. In fact, police access to such externalized thought may, in many cases, make their need for surveillance of internal thought processes unnecessary. As Orin Kerr has recently written, the shape of Fifth Amendment law, for example, in encryption cases, may often depend on how often government can obtain the evidence crucial to understanding the events that will be at issue in a trial: "If strong encryption is a big barrier that substantially impacts a lot of cases, pressure for change will mount; if it turns out the government can still solve many cases anyway, the pressure will dissipate" (Kerr 2016). The same may be true in context of neuroimaging: Access to externalized thought may allow the justice system to function without accessing our unexpressed thoughts.

One might argue that this kind of "SmartPhone- and Computer-first" regime gets privacy-protection priorities backwards – that the details of life within our SmartPhone or computer files are likely to be more revealing than the information about our mental states that police could obtain from any kind of brain imaging likely to be available to them in the near future. As a consequence, one might argue, brain imaging should be easier for police to access, for example, than it should be for them to access our digital files. But the realm of unexpressed thoughts is a realm where individuals have traditionally expected greater privacy. They have assumed, for example, that while phones or computers could be hacked, the plans or images they have made only in their "mind's eye" are inaccessible even with advanced technology. And seems intuitively unlikely that people would view a hacking of one's phone (or destruction of it) to intrude into one's "actual 'self'," to the same extent as an investigatory method that monitors and records one's brain activity. Such intuitions about privacy matters certainly matter in Fourth Amendment law, and it is worth considering the possibility that they may also matter to self-incrimination doctrine.

CHAPTER 5

The Fourth (and First) Amendment: Searches with, and Scrutiny of, Neuroimaging

Abstract The questions raised of Fourth Amendment law by neuroimaging at first seem to have simple answers: The Fourth Amendment covers neuroimaging because probing any part of the body's interior is a "search." The standard level of protection against such a search is the warrant requirement, imposing on government the responsibility of showing probable cause and specifying the place to be searched before conducting such a search. However, matters are not so simple. There is significant gray area in the Fourth Amendment that the court has used to give government flexibility in meeting vital security interests. This chapter shows that some of the answers to these Fourth Amendment problems may unexpectedly have First Amendment solutions.

Keywords First Amendment · Fourth Amendment · Free speech · Neuroimaging · Search · Searches incident to arrest · Special needs searches · Third party doctrine · Warrantless · Warrants

An Overview: Why Neuroimaging Raises Fourth Amendment Problems

In one key respect, the Fourth Amendment questions raised by neuroimaging are far easier to answer than those raised by the Fifth Amendment's self-incrimination clause. Does the self-incrimination clause bar compelled

© The Author(s) 2017
M.J. Blitz, *Searching Minds by Scanning Brains*,
Palgrave Studies in Law, Neuroscience, and Human Behavior,
DOI 10.1007/978-3-319-50004-1_5

neuroimaging of a criminal defendant? The answer isn't clear because it is not clear if neuroimaging represents the kind of "self-incrimination" covered by that clause. As explained in the previous chapter, it is only incrimination through a witness's testimonial evidence – not through physical evidence – that is covered by the clause. So, we have to determine on which side of this dividing line neuroimaging falls.

By contrast, our task is easier when we ask if the Fourth Amendment constraints compelled neuroimaging. It constrains any police investigatory technique that counts as a "search" or "seizure." Is a brain scan a "search"? As I noted in Chapter 2, the answer is almost certainly "yes." And almost all scholars who have addressed the question agree it is "yes." A government engages in a search when it intrudes upon an area where one has a "reasonable expectation of privacy," (Katz v. United States 1967, 360–361) and the interior of our bodies is one such realm. As Michael Pardo writes, "subjects have a 'reasonable expectation of privacy' in information about their brain states," just as they do about "other information about inner bodily processes such as the contents of one's blood or urine" (Pardo, 2006, 325). Police certainly intrude upon an individual's privacy, as well as his dignity and comfort, when they take a blood test which, as the Court has observed, requires "piercing the skin" with a needle and "extract[ing] a part of the subject's body" (Skinner v. Ry. Labor Executives' Ass'n 1989, 625). Urine tests do not require such surgical "intrusion into the body," but as the Court observed, they are nevertheless searches since they "can reveal a host of private medical facts about an employee" and because "visual or aural monitoring of the act of urination, itself implicates privacy interests." Breath-testing procedures too, said the Court, "implicat[e] similar concerns about bodily integrity and, like [a] blood-alcohol test should also be deemed a search" (Skinner v. Ry. Labor Executives' Ass'n 1989, 616).

Brain scans are, in some respects, less intrusive than all of these other searches of the body, since they do not require the individual to provide authorities with any biological material. Authorities merely observe brain activity – and obtain information from it. Still, even a medical device that simply extracts information from inside the body counts as a search. As noted earlier, courts have consistently held that where the government uses magnetometers or X-ray devices at airports to reveal information about what a traveler is carrying beneath a jacket, or inside a bag, this is a "search" – even if the authorities do not physically examine the traveller's clothing or bag (United States v. Epperson 1972, 770; United States v. Albarado 1974, 803–805). It seems clear that it is likewise a search when authorities use electrodes, radio waves, near infrared radiation to probe more deeply – and gather information not just from underneath her clothing, but underneath her skin. Indeed, while these X-ray and magnetometer cases have been decided by federal circuit courts, the Supreme Court itself has made it clear that law enforcement personnel cannot use advanced technology that lets them see or listen through walls to violate the home's privacy and integrity, even though police can thus search a home from a public place. They cannot, said the Court, use a radio transmitter to track a person's movements inside his home (United States v. Karo 1984, 714–718). Nor can they use a thermal imaging device to collect details about a home's interior from a public street outside (Kyllo v. United States 2000, 34). As Pardo notes, if, as the Court has made clear, "one has a reasonable expectation of privacy in the details of one's home (even when measured from outside with a thermal-imaging device)…one plainly also has a reasonable expectation of privacy in the details of what is in her head, even though the government doesn't have to invade the body to learn the information" (Pardo 2006, 325). Madison Kilbride and Jason Iuliano similarly argue that "[j]ust as searching a person's house with a thermal-imaging device or eavesdropping upon a person's phone conversations undermines that individual's privacy interests without invading his bodily space, so, too, does neuro lie detection infringe upon a person's right to privacy in a non-physical manner" (Kilbride and Iuliano 2015, 141–42). Amanda Pustilnik likewise argues that compelled neuroimaging should count as a "searc[h] of private information within a space of presumed privacy," and thus receive Fourth Amendment protection (Pustilnik 2013, 121).

The more complex Fourth Amendment issue raised by neuroimaging technology is not whether US citizens are protected from law enforcement use of it (they are), but how much protection they get. This may initially seem like an easy question too. Typically, when a particular investigatory measure is a "search," police cannot carry it out unless they first obtain a warrant from a neutral magistrate. To do so, they have to specify the "place to be searched" and explain why they have "probable cause" to believe they will find evidence of a crime where they wish to search. So, one might assume that if police wish to obtain a brain scan from a person, that is precisely what they will have to do. They will have to specify whose brain it is that they wish to scan. They will then have to explain to a neutral magistrate why they expect that a brain scan will reveal evidence related to criminal activity (for example, by revealing that the person recognizes the murder weapon or by allow them to infer some other memory about how the crime was committed). If the magistrate is convinced by these arguments that police have "probable cause" to conduct the search, she will issue a warrant permitting them to do so, and defining its scope.

So, in sum, there is a fairly straightforward answer to what many scholars call the "coverage" question raised by neuroimaging: It is a "search" and thus covered by the Fourth Amendment's reasonableness constraints. There is also a fairly straightforward answer to the "protection" or "procedure" question: Police can typically conduct a search only if they have a warrant based upon probable cause and particularly stating where they will search. In fact, in the normal circumstance, a Fourth Amendment search carried out without a warrant is "per se unreasonable" – and thus, unconstitutional (Katz v. United States 1967, 357).

Why then is there any Fourth Amendment puzzle about whether and when police can use neuroimaging? Such puzzles arise from the extraordinary amount of gray area within Fourth Amendment law itself – a gray area that the US Supreme Court has had to wrestle with practically every year during the past decade as it tries to figure out how emerging technologies – raising unprecedented challenges to privacy – fit into Fourth Amendment doctrine, and whether they demand that this doctrine be changed.

Let me preview some of the key uncertainties about neuroimaging's Fourth Amendment status. First, two complications in coverage: The government engages in a Fourth Amendment search when it requires citizens to, say, submit to an EEG test. Would it be doing so, however, if without revealing the law enforcement use they would make of neuroimaging data officials convinced someone into unwittingly providing such data on a brain-computer interface for a video game, or that one uses in surfing the World Wide Web? Or if they obtained such EEG data from a non-governmental hacker (perhaps a professional "brain hacker")? Under the so-called "third party doctrine," one might argue that it wouldn't be. Under that doctrine, when we share our information with a third party, we assume the risk they will pass it on to government (voluntarily or in response to a subpoena). And if that risk materializes, it's still not a Fourth Amendment search. Under existing Fourth Amendment doctrine, we can't claim that government has invaded our private space to obtain such EEG information – when it obtained that information not from a private space but rather from another source to which we willingly provided it.

There is also another development (albeit an unlikely one) that could make Fourth Amendment coverage vanish. Sometimes, even a government probing of a highly-private space is not a "search" – because the constraints the Fourth Amendment is designed to impose are, in effect, already built into the technology. One of the few examples is a "dog sniff" by a police canine trained to alert only to the presence of cocaine or heroin. If police bring the dog near our car or our suitcase, they are in effect exploring the interior of these private realms – something that would normally be a search. But the dog can only provide one piece of information about this private interior realm – and that is whether it contains illegal drugs. And that piece of information about what lies within a private interior space is something, the Court has found, in which we have no legitimate privacy interest. One might ask then: Are there certain kinds of neuroimaging that would reveal only memories or knowledge that a person has no right to keep private – for example, a memory that reveals the person responsible for a murder or a kidnapping? If so, could that make such use of neuroimaging a non-search, especially if the testing technology told law enforcement nothing else about the brain activity of the person being tested. For reasons I will explain, this argument is unlikely to work: Even a breathalyzer test that tells a police officer nothing but a person's blood alcohol level still involves an intrusion into bodily privacy of a sort that a dog sniff of luggage or a car does not. And the same is likely to be true of neuroimaging. Still, it is useful to raise this question, both because it helps us think a little bit more about the Fourth Amendment interests at stake when someone is subject to neuroimaging, and also because it helps lay the groundwork for certain questions that arise when we turn to Fourth Amendment protection.

And neuroimaging also raises questions about such protection. Indeed, most of the Fourth Amendment gray area that is relevant to neuroimaging concerns the question of "protection." I will say a little bit about these questions here and then look at them more closely later. Consider, first, some of the questions that might arise even if it is clear that a warrant is required. More specifically, imagine that a group of law enforcement investigators wishes to take a brain scan of a person whom they have probable cause to believe is involved in a counterfeiting operation. They also have suspicions that the same person may be involved in a plan to carry out domestic terrorism in the coming months, and that money from the counterfeiting operation may be directed to those plans. On the basis of information suggesting a link between this person and the counterfeiting, they seek a warrant from a judge to take a brain scan of the suspect so they can run brain scans to gather more detailed information about the operations and his connection to it. One set of questions likely to arise concerns the scope of the warrant: What will such investigators have to tell a judge to "particularly describe the place to be searched"? Is it enough for them specify that they are seeking evidence about a counterfeiting operation and will try to derive it from using an EEG or fMRI to probe the suspect's response to particular stimuli? Or do they have to describe the methods to the judge in more detail, for example, by describing what kind of images, words, or other stimuli they will show the suspect, so it is clear they will not gather more information about his beliefs, thoughts and feelings than what they need? If, in gathering information about the counterfeiting operation, they find evidence which they believe is relevant to the terrorism plans they believe to be associated with it, or evidence of other criminal activity, do they need another warrant to capture and analyze such additional neuroimaging evidence?

Moreover, one might ask, should a warrant that authorizes a brain scan require nothing more than probable cause and satisfaction of the particularity requirements? Or when mental privacy is at stake in this way, should the government be required to meet a higher threshold? Should it, for example, be required to show (as it is when it seeks a warrant to conduct surreptitious video surveillance in a private area) that other less intrusive measures of obtaining the same information won't work?

Other Fourth Amendment questions will arise if the government wants to use neuroimaging technology in circumstances that have, in the past, normally made it "reasonable" and thus permissible for government to use a warrantless – or even suspicionless – search. The government, for example, does not need a warrant to search you as you enter the United States. It doesn't even need reason to suspect you. There is a "border search" exception to the warrant requirement. The same is true at airports. No matter how unlikely it is that you are a threat to airline safety, if you are a traveler planning to board a flight, you need to go through a magnetometer or other screening device, and submit your luggage to similar screening (and to any physical searches the government deems necessary). Government may also conduct warrantless searches "incident to arrest": it can, for example, require individuals arrested for drunk driving to submit to breathalyzer tests to determine the level of alcohol in their bodies.

How, one might ask, might neuroimaging fit into each of circumstances? Would a permissible warrantless search remain permissible if the government decided to add brain scanning to its use of magnetometers or breathalyzers? Or would that kind of evidence-gathering about brain operations and mental activity bring it to a Fourth Amendment red line which – even at a border station, an airport, or a police station processing an arrest – it can't permissibly cross without careful judicial oversight? May police, for example, subject an arrestee to a brain scan shortly after arresting him in order to see if any of his accomplices are nearby? May they try to gain some access to knowledge of his accomplices' whereabouts before it becomes stale, or he begins to forget? Might school officials worried about the threat of school violence, or airport officials worried about terrorism, add neuroimaging to the set of tools they use to detect and thwart violent threats?

Fourth Amendment Coverage – and Kyllo V. United States

As Orin Kerr writes, Fourth Amendment doctrine, and its rules of coverage, effectively divide up government evidence gathering into two parts: There are "less invasive steps the government can take at any time, and more invasive steps the government can only take when it has already collected enough evidence to demonstrate special conditions such as probable cause" (Kerr 2009, 574). For example, when police monitor events in public space – for example, a "park or on open fields" – they are conducting surveillance that is unlikely to interfere with our private lives and which they can therefore undertake whenever they like. By contrast, when police want to look into a "home or private packages" – they *are* conducting a Fourth Amendment search and therefore need to obtain a warrant or otherwise show their investigation is constitutionally reasonable. This line between "less invasive" steps in public space and "more invasive steps" into homes or other private spaces is essentially the line that marks the boundaries of Fourth Amendment coverage: The private spaces are covered by the Amendment, the public spaces are not.

This framework provides yet another way to reaffirm the scholarly consensus on neuroimaging: It gathers evidence from inside our bodies, so it is among the "more invasive" steps that law enforcement officials can only conduct if they satisfy Fourth Amendment reasonable require-ments, typically by first obtaining a warrant. But before we treat that as the full story on Fourth Amendment coverage for brain imaging, it is useful to say more about this framework – and it is useful to use a 2001 case called Kyllo v. United States to better understand how the Fourth Amendment divides up the world into (1) spaces where the government has free reign to gather evidence, and (2) spaces where we can generally exclude it, and prevent evidence-gathering, unless and until government has good justification for entering.

Kyllo is a case that receives mention from neuroscience and law scholars for a number of reasons. Pardo, as noted earlier, cites it to show that one can conduct a search of the body (as of the home) even when standing outside it and using technology to peer in (Pardo 2006, 325). Amanda Pustilnik cites it both for that reason – and also to argue that, just as the Court in Kyllo was protecting the privacy of the life within the home, and not simply the home as a structure, so Fourth Amendment protection for neuroimaging might protect not just the privacy of our brain physiology, but the interior mental experience that it generates (Pustilnik, 2013, 131–134).

Kyllo essentially held that when police use a thermal imager to gain information about a home's interior, they are conducting a Fourth Amendment search – and need a search warrant to do so. An agent of the US Department of Interior, William Elliott, suspected that an individual by the name of Danny Kyllo was growing marijuana inside his home. Such in-home marijuana cultivation required use of a "high intensity lamp," such as a halide light, and Elliott and a colleague realized that, if Kyllo were using such a lamp, it would – because of its heat, emit plenty of infrared radiation – radiation they could detect from outside of Kyllo's home with a thermal imaging device. They therefore pointed such a device at Kyllo's home, while sitting in their car across the street from it, and found that an area over the garage of the home emitted considerably more infrared radiation than the rest of the home, and considerably more than was emitted from the homes of Kyllo's neighbors (Kyllo v. United States 2000, 30). On the basis of this information, as well as information in utility bills (to provide more evidence that an energy-intensive lamp was being used inside) and tips from informants, the agents obtained a warrant to enter and search Kyllo's house – where they found the lamp and "in indoor growing operation involving more than 100 plants" (Kyllo v. United States 2000, 30).

Kyllo, however, argued that the agents needed to obtain a warrant – based upon probable cause – not only to enter and search his home, in the later stage of their investigation, but also before they pointed the thermal imager at his house, and recorded the infrared radiation that could give them clues as to Kyllo's private activities. Pointing the thermal imager at the home, he argued, was itself a Fourth Amendment search (Kyllo v. United States 2000, 37).

The Supreme Court agreed. Agents could, it acknowledged, look at a home from the street outside – and such visual surveillance by itself would not constitute a search (Kyllo v. United States 2000, 32). So, for example, had Elliott and his colleague stood across the street from Kyllo's home, stared at it, and seen bright light from the halide lamp through a large open window, they would not have been engaged in a search. Nor would they have engaged in a search if they conducted such visual surveillance, and noticed drug paraphernalia lying in a garbage can in the street outside Kyllo's home. In fact, the Court had previously ruled that even if agents hovered over a greenhouse in a helicopter, and noticed marijuana being grown through a crack in the greenhouse roof, such visual surveillance from an aerial perspective wouldn't be a search either (Florida v. Riley 1989, 450–452). But the investigation of Kyllo's house with a thermal imaging device, said the Court, involved "more than naked-eye surveillance of a home" (Kyllo v. United States 2000, 33). The device allowed police to perceive activities within the home's interior not through an open window, or a crack in its structure, but through a solid wall. That insider's perspective on the home, reasoned the court, would previously have been impossible without entering the home itself. And when "sense enhancing technology" is in this way a functional substitute for "physical intrusion into a constitutionally protected area," then use of it by law enforcement should be subject to the Fourth Amendment, and subject to its privacy safeguards (Kyllo v. United States 2000, 34).

Pustilnik offers Kyllo as a good model for explaining Fourth Amendment coverage for an EEG measurement: Just like "thermal signatures from a home," she explains, "electrical brain waves are automatically and continuously produced," and so one might argue that they are – in some sense – there for the taking for anyone who can unobtrusively gather such information from outside (of a private home, or of person's body) (Pustilnik 2013, 133). But both of them are not only "invisible and undetectable absent technology" – they can also both be "decod[ed]" to reveal information that people typically consider to be at the core of their privacy: The home life people are able to shield from public view and thoughts they avoid sharing (sometimes even with family or others in their home) (Pustilnik 2013, 133). Moreover, the Court also took a position that is instructive for analysis of brain scan technology by focusing not only on the crude capacities of thermal imaging existing in the late 1990s (when Kyllo's home was searched) but also on the "imaging technology" that might develop in the future and might "discern all human activity in the home" (Kyllo v. United States 2000, 35). In applying Fourth Amendment law to brain scanning, courts might likewise take account not just of brain scanning as it exists today, but also as it is likely to evolve over the coming decades.

Kyllo's template for covering new surveillance technologies is certainly an attractive and useful one. But we need to say more about it – and how brain scan technology might be analogized to it – in order to understand whether it provides the basis for an approach to Fourth Amendment coverage that takes sufficient account of the value of intellectual privacy. More specifically, it is helpful to understand Kyllo not only by looking at the Court's conclusion and the core reasons it gives – but at the other side of the debate that occurred between different Justices in that case. While Justice Antonin Scalia held, supported by a majority of the Court, that police engage in a search when they use a thermal imager to gather information from the home's interior, Justice John Paul Stevens vigorously argued that it was not a search – and the Fourth Amendment should thus leave such police use of thermal imaging unconstrained. This was not because Stevens believed that the privacy of the home was unimportant. On the contrary, he emphasized in his dissenting opinion that "the homeowner has a reasonable expectation of privacy concerning what takes place within the home, and the Fourth Amendment's protection against physical invasions of the home should apply to their functional equivalent" (Kyllo v. United States 2000, 44). But this, he said, could not shield from police view, what was visible on the outside of the home's walls. The home's exterior walls and the environments outside of the walls were part of the "public domain," that – in a free society – can be viewed by citizens other than the property owner (including police officers). Thus, he noted, "any member of the public" on the street outside the house "might notice that one part of a house is warmer than another part or a nearby building if, for example, rainwater evaporates or snow melts at different rates across its surfaces" (Kyllo v. United States 2000, 43). Or he might make inferences about what a person is cooking inside a kitchen by smelling the aromas that float outside (Kyllo v. United States 2000, 43). The thermal imager, said Stevens, simply took as its raw material, and then refined, such physical occurrences in the home's exterior – in this case, the heat that leaked outside of its walls.

We might elaborate and sharpen Justice Stevens's objection here by connecting it with an observation the Supreme Court made in an earlier case where it found that police could gather certain information outside of the home from a public vantage point: In Ciraolo v. California, the Court held that police did not conduct a Fourth Amendment search when, in following up an anonymous tip that a home owner was growing marijuana in a backyard garden, they flew a plane 1000 feet over the garden and identified the marijuana plants below (Ciraolo v. California 1986, 209). In explaining why it was not a Fourth Amendment search for police to conduct such a fly-over, the Court emphasized that this kind of observation from a public vantage point is "precisely" the kind of information that "a judicial officer needs to provide a basis for a warrant" (Ciraolo v. California 1986, 213). In other words, our homes can remain constitutionally insulated from police investigation only because there is some *other* place – namely, the public domain – where police can vigorously gather evidence of possible criminal activity before taking the more extraordinary and intrusive step of entering the home. If even the large swathes of public domain become off-limits to warrantless police observations then they will have nowhere to begin their investigations. To obtain a warrant allowing them to search the inside of a home for marijuana operations, they need to first get some information from somewhere outside of a home to show that such an inside search is justified. If the law then tells them they need a warrant to scrutinize the outside of the home (with, for example, aerial surveillance or a thermal imaging analysis of heat emissions), then it may be hard to see how (and where) police can begin to build a case for focusing on a particular target. Justice Stevens's dissent then, can be read as an argument that while a just balance between crime fighting and privacy can bar police from closely examining the details of the home's interior – it cannot also reasonably demand that police ignore clues about this interior activity that they find in the public domain. By preventing police from learning anything about a home's interior, even from evidence available in the street outside, one will be making it impossible to tell when the home is being used not simply as a refuge in which people can find privacy from the world, but also as a site for hiding criminal operations.

Stevens's dissent also contains another important challenge to the majority decision in Kyllo. That decision had emphasized that when law enforcement aims to, and succeeds, in extracting information from the home's interior, it does not matter how personal or sensitive that information. Even information that seems so mundane as to be non-private is still staunchly protected from police observation when it is inside the home: as the Court says, entry into the home constitutes a Fourth Amendment search requiring a warrant even if an officer "barely cracks open the front door and sees nothing but the nonintimate rug on the vestibule floor . . . in the home . . . all details are intimate details, because the entire area is held safe from prying government eyes" (Kyllo v. United States 2000, 37). But if one is prevented from learning non-intimate details, even with technology employed outside the home and focused only on the home's exterior or its environment, then – Stevens worries – law enforcement may be barred even from common-sense strategies for "detect[ing] the odor of deadly bacteria or chemicals for making a new type of high explosive" (Kyllo v. United States 2000, 48). Even where a technology is – "like [ ] dog sniffs" designed to detect only drugs or explosives – or to detect nothing more than contraband or dangerous weapons – the majority's rule in Kyllo, worries Stevens, may still make it impermissible for police to use it without a warrant. In other words, not only do police have a right to investigation (without any warrants) outside the home. They also have a right, on Justice Stevens's view, to try to gather information even from inside the home, when the only information they are trying to gather is information of the kind a person has no right to keep private – such as the presence of "chemicals for making a new type of explosive" (Kyllo v. United States 2000, 48).

None of this is to say that Justice Stevens or any other Supreme Court justice would have excluded brain-based mind reading from the Fourth Amendment's coverage – or ignored the privacy interests that may be at stake in it. On the contrary, Stevens made it clear that while he believed that it is not a search for police to use a thermal imager to detect heat emanations on the exterior, it *would* be a search if they used what, in his view, was the electronic "functional equivalent" of a "physical invasion of the home." The agents who viewed Kyllo's home, for example, would have conducted a search, argued Stevens, had they been able to use something more like "an x-ray scan, or other possible 'through-the-wall' technique" (Kyllo v. United States 2000, 43). An fMRI or fNIR brain scan derives evidence of brain processes by gathering information from blood flow inside the body. It does not, like the reading of micro-facial gestures, seek to detect dishonesty or other internal mental states from observations of a person's visible behavior. And Stevens would also probably be amenable to the argument that what makes our invisible brain processes so private is not simply that they are inside the body, but that they can be used to infer evidence of our thoughts.

Still, Stevens's dissent in Kyllo – and the similar arguments the Court has made over the years for assuring that police are left with some space in which they can find the necessary evidence to lay the groundwork for successful crime-solving work – raise important questions about the basis, and extent, of Fourth Amendment coverage for brain scanning. In short, Stevens's arguments emphasize that there are at least two situations where Fourth Amendment coverage that is normally present – say for activities in the interior of a house – vanishes: (1) where police do not collect evidence from inside the house, but collect information from the public world outside of the home – either because they can observe it there, or because they can obtain it from someone who shares it with them, or (2) where instead of remaining outside the home, police use a technique that penetrates into the interior but tells them only about activities that the individual has no right to conduct – such as possessing illegal drugs or manufacturing explosive materials. Each of these highlight two possible exceptions to the general rule that brain imaging would almost certainly be a Fourth Amendment search and it is worth considering each of them a little more closely.

FOURTH AMENDMENT COVERAGE – AND THIRD PARTY DOCTRINE AND ABANDONMENT

There are certain activities we take that we do not leave entirely open to observation, but which the Supreme Court has long said are public enough to be fair game for government to investigate without a warrant, and indeed, without any Fourth Amendment limits. This happens, for example, when we abandon items, leaving them to be examined by any one who recovers it (including police). For example, when we dispose of garbage, we often enclose it a sealed bag: We do not leave it open to observation by the rest of the world. But because we are disclaiming any property interest in it and leaving it to be picked up by a trash service, we lose the right to treat it as a private space under our control.

This abandonment doctrine was central to the Court's ruling in the 1988 case Greenwood v. California. In that case, police had received tips that a man named Greenwood was engaged in drug trafficking. So, they contacted the trash collector for the neighborhood, and asked him to preserve Greenwood's trash for police to look at, making sure that it wasn't mixed with those of others in the neighbourhood. When police then examined the bags they received from the trash collected, they found items indicative of drug use – and used them to obtain a warrant to search Greenwood's house. Greenwood claimed (as Kyllo would later do) that they needed a warrant not only to search the house, but even earlier in their investigation – in this case, when they wanted to search his trash. The Court, however, held that he lacked a reasonable expectation of privacy in its content: Once left on "a public street" where it was accessible to "animals, children, scavengers, snoops, and other members of the public," he could retain no further interest in it. Having abandoned the items in the garbage, Greenwood could not expect to control who examined them (Greenwood v. California 1988, 40–42). This was true, even though there are many things people dispose of that they might be uncomfortable sharing: As the dissenting opinion pointed out, "A single bag of trash testifies eloquently to the eating, reading, and recreational habits of the person who produced it. A search of trash, like a search of the bedroom, can relate intimate details about sexual practices, health, and personal hygiene" (Greenwood v. California 1988, 50).

As Elizabeth Joh points out, courts analyzing "abandoned DNA" have followed Greenwood's analysis and have concluded that "there is no objective expectation of privacy in saliva – and the DNA contained within it – that is left behind on a coffee cup or on a smoked cigarette." Indeed, in some circumstances, suspects have been tricked into providing DNA samples. Police have sometimes "act[ed] as passive collectors, waiting for a suspect to discard a smoked cigarette or to spit on the floor" (Joh, 2006, 868–872). In one notable use of a similar DNA-collection technique in a Washington murder investigation, "Seattle Police Department detectives, posing as a fictitious law firm, induced [the suspect,] Athan to mail a letter to the firm, from which Athan's DNA sample was extracted." Following the pattern described by Joh, the Washington State Court found in that case that, because "obtaining the saliva sample in this case did not involve an invasive or involuntary procedure," Athan retained no privacy interest in the saliva he voluntarily used to seal the envelope (State v. Athan, 2007, 31, 33).This abandonment doctrine has been subject to scholarly criticism, by Joh and others. For example, in analyzing Canada's equivalent of Fourth Amendment protection, Ian Kerr and Jena McGill explain that if it is constitutionally problematic for police to learn about the interior of our homes from heat "emanations" released from it into the outside world, it should be similarly problematic for police to learn about a person's interior through use of the information "our bodies emanate" in "DNA from flaking skin cells and shedding hair" and "electrical activity from brains and hearts" (Kerr and McGill, 2007, 393). But Fourth Amendment law continues to treat the latter type of information sources as unprotected.

There is also another circumstance where information is treated as open to police observation even when it is public: When we convey information to a third party, that party can then share it with government (whether voluntarily or in response to a subpoena), without that sharing counting as a Fourth Amendment search. This doctrine first originated in cases involving government informants: In On Lee v. United States, the government recruited a friend of a suspect drug dealer to act as an informant and surreptitiously record the suspect with a hidden microphone (On Lee 1952, 748–751). Similarly, in Hoffa v. United States, Jimmy Hoffa made the mistake of confiding in a friend who was secretly working with the police (United States v. Hoffa 1966, 294–302). In these and other cases, the Court made clear that the government's use of an informant was not the kind of invasive technique that counted as a Fourth Amendment search: The Fourth Amendment, said the Court in Hoffa, does not protect individuals against misplaced trust.

The Court then extended this logic to make clear that other kinds of sharing of information we undertake are also taken at our own risk. Even if we are dealing not with friends (who we might or might not decide to continue trusting), but rather with companies we have little choice but to deal with if we want to participate in modern life – companies such as banks or telephone companies – we still lose the right to prevent these third parties from, in turn, sharing it with government. Thus, in Smith v. Maryland, the Court held that it is not a Fourth Amendment search for law enforcement to obtain, from the phone company, records of the phone numbers dialed from a customer's home (Smith v, Maryland 1979, 741–743). In United States v. Miller, it likewise held that it is not a Fourth Amendment search for law enforcement to obtain, from a bank, copies of checks an individual has deposited. While we may hope and expect that these businesses will maintain the privacy of our calling or financial records, our sharing of this information nonetheless places it into the realm of information where police's access to it is considered a "less invasive" investigative step, not subject to Fourth Amendment limits (United States v. Miller 1976, 441–443).

One question we might ask about brain imaging data then is whether there might be circumstances where it too is treated as abandoned or shared in a way that takes it outside of the Fourth Amendment's ambit. This is certainly not the scenario most writers imagine when they imagine law enforcement use of brain imaging: They imagine a government agent asking questions of a subject undergoing a brain scan, or presenting him with images or other stimuli and assessing his brain responses. But there are three scenarios – further removed from existing technology – where third-party doctrine would be more relevant.

First, as Stacey Tovino observes, insurance companies, for example, could conceivably want "individuals' neuroimaging information…to predict future illness, a propensity to violence, or other conditions or characteristics relevant to underwriting decisions" (Tovino 2005, 847–848). Employers may also have an interest in collecting neuroimaging information about their employees. As Tovino notes "the federal Employee Polygraph Protection Act ('EPPA') prohibits employers from requiring employees to submit to lie-detector tests, defined to include polygraphs, deceptographs, voice stress analyzers, psychological stress evaluators, and 'any other similar device'" and this would likely cover use of brain imaging to assess an employee's honesty. But this leaves open the possibility that neuroimaging information would be used for other purposes. And it is conceivable that individuals seeking certain types of psychiatric or other medical treatment would undergo brain imaging, leaving records that might be claimed by government.

Of course, if neuroimaging measures only individuals' responses to specific stimuli, then it is unlikely it will include information of precisely the kind law enforcement is looking for in solving a specific crime. An insurance company seeking to assess an individual's tendency to take unwise risks will not typically have reason to ask whether a client is familiar with aspects of a particular crime under investigation.

Second, apart from situations where companies collect neuroimaging information from individuals, and then share it with law enforcement, it is conceivable that we will collect neuroimaging information about ourselves – or leave a data trail about it as we play video games we control with EEG technology or with other kinds of brain-computer interfaces. As noted earlier, companies such as Emotiv and Neurosky have marketed video games that use a variation of EEG technology to allow gamers to control video-game play (Childers 2013). In 2012, Ivan Martinovic and his colleagues conducted experiments demonstrating "the feasibility of using a cheap consumer-level BCI gaming device to partially reveal private and secret information of the users." They attempted to use evidence – from EEG-based gaming systems – of P300 reactions to infer information about banks and credit cards (Martinovic et al. 2012). They also noted that while their own experiment was fairly simple, it was easy to conceive of "more sophisticated attacks" – for example, attacks in which "an uninformed user could be easily engaged into 'mind-games' that camouflage the interrogation of the user and make them more cooperative." Martinovic and his colleagues were primarily concerned about how mental privacy could be compromised in identity theft or other crime. But techniques that "camouflouge the interrogation of a user" could also, of course, be used by a government informant, like those in On Lee and Hoffa. Tamara Bonaci and Howard J. Chizek also analyze the privacy concerns that arise as individuals use brain-computer interfaces – and note that "several marketing companies . . . have shown interest in the usage of BCI devices for marketing research" and that "if BCI devices become widespread we might see private information extracted from individuals without their permission" (Bonaci & Chizek 2013).

As a number of writers point out, intellectual privacy has already been compromised by the migration of numerous activities – including those involved in intellectual exploration – to the Internet. Julie Cohen, for example, pointed out in 1996 that individuals' reading habits, which had been a traditionally private activity, were increasingly subject to being monitored as people switched from reading physical books to digital books on the Web (Cohen 1996, 1004–1019). The spread of Kindle and other electronic reading services makes it easier to track reading habits. More recently, Neil Richards has pointed out that an "your ISP has records of every website you've visited – transcript of your intellectual explorations, of your reading and thinking" (Richards 2015, 5). To the extent that brain computer interface (BCI) interactions merge with on-line gaming or other interactions, we may add to this trail of data about our reading and Web-searching choices, data that might reveal some of the feelings or other internal reactions we have as we interact with Web-based content.

Third and finally, there is one more scenario that might make it relatively simple, and common, for parties other than government to obtain information that they can share with government agents. Nita Farahany notes that "police may soon be able to monitor suspicious brain activity from afar" and that "various government agencies are funding the development of technology to detect brain activity remotely and are hoping to eventually decode what someone is thinking" (Farahany 2008). Of course, even if government can monitor the brain from a distance, it would likely be just as engaged in a Fourth Amendment search as it is when it detects a home's infrared radiation from a decent distance. But if technology arises that allows government to monitor a "suspicious brain from afar," the same technology will allow others to conduct such monitoring – and then share such information with government, free of Fourth Amendment restraints.

Indeed, if such technology for remote brain imaging became sufficiently widespread – use of it may lose its status as a Fourth Amendment
search for another reason. In Kyllo, even Justice Scalia, in noting that the
Fourth Amendment should protect individuals from future as well as
present technology, implied that the thermal imaging that was a search
in that case might not be in the future. If and when thermal imagers ceased
to be an uncommon and relatively unknown tool police could use to see
into places where individuals expected to remain unseen – if they instead
became a technology in "general public use" that Americans knew were in
others' hands, and understood might affect their privacy – then police, he
hinted, might be as free to use them as anyone else (Kyllo v. United States
2000, 34). Other scholars have pointed out that it is unclear how the court
would determine when an investigatory technology is in "general public
use" (Slobogin 2007, 57–58, 62–65, Adkins 2002, 262). However, it is at
least possible that if neuroimaging devices become perhaps as pervasive as
SmartPhones are today, and individuals can (legally) use them not only to
gather information about their own minds, but also those of their neighbors, then use of such technology may no longer count as a Fourth
Amendment search – even when it gathers information from within a
person's body (Federspiel, 2008, 881–882).

In short, while the third-party doctrine is unlikely to leave police with
free access to any kind of brain imaging data in the near future, it may be a
part of the Fourth Amendment analysis in future technological landscapes
if the technology both migrates out of the laboratory, and is used by
individuals and private companies. This does not mean, however, that
the Fourth Amendment doctrine of the future will necessarily leave government with access to such data.

Third-party doctrine has long been harshly criticized by Fourth
Amendment scholars. Daniel Solove for example, argues that "[g]overn-
ment access to records held by third parties should be covered by the
Fourth Amendment" (Solove 2010, 1533). Stephen Henderson, writes
that "the doctrine was controversial when adopted, has been the target of
sustained criticism, and is the predominant reason that" in the view of
many scholars Fourth Amendment law still offers far too little protection
in the face of technological change (Henderson 2007, 1976). He notes, in
a more recent article, that the third-party doctrine left the Fourth
Amendment ill-equipped to protect us against automatic sharing of information and that courts seemed to recognize this: "Fourth Amendment
Third Party Doctrine – which holds that a person retains no expectation of

privacy in information conveyed to another – has at least taken ill, and it can be hoped it is an illness from which it will never recover," and he proposes a four factor test that could offer Fourth Amendment protection to at least some kinds of third-party records – one factor of which protects transfer of records linked to First Amendment freedoms (Henderson 2011, 39–40, 48). Jane Bambauer similarly argues that "[t]he third-party doctrine may be dismantled soon, and for good reason. It always strained the logic and common sense of search and seizure law" (Bambauer 2015, 208). One reason for these predictions that third-party doctrine may not survive coming years of technological change is the opinion of Justice Sotomayor in United States v. Jones, which – although focused on the question central to the case about whether and when police can permissibly use GPS technology to track a car's movements in public – added that "it may be necessary to reconsider the premise that an individual has no reasonable expectation of privacy in information voluntarily disclosed to third parties" (United States v. Jones 2012, 975).

However, as Henderson writes, it is one thing to reject the doctrine and urge that "some third-party information should be protected, and quite another to articulate how and when different information should be accessible to police" (Henderson 2007, 976). Whether brain imaging information is protected, even when we let it be captured in computer interactions or kept by third parties, would depend on what specific criteria courts adopted to decide what third-party records merit Fourth Amendment protection. What, in other words, might place certain third-party records back within the realm of the Fourth Amendment – on the private side of the private-public distinction, such that law enforcement access to and use of those records would be a "more invasive" method constrained by the Fourth Amendment rather than a "less invasive" one free of its restraints?

One argument, which I will elaborate upon in the next section, is that the relationship of a record to our intellectual privacy makes a difference here. Solove already makes an argument like this: Where third-party records have First Amendment value, at least some constitutional provision (the First Amendment if not the Fourth) should safeguard them: Such records would include "bookstore records" which "clearly fall within the boundaries of the First Amendment because they concern the consumption of ideas." "Internet search queries," he argues "are very similar to book records in that they involve a person's reading habits and intellectual pursuits" (Solove 2007, 286).

Data that allows authorities to draw inferences about our thoughts should similarly be among the kind of records that receive heightened protection against government monitoring. Even fragmentary evidence about our thoughts – the fact that we recognize an image, or feel anxious in response to it – should not be evidence government can help itself to when our response is manifested not in a visible observation (like a facial expression) but rather in the data generated by our use of a neurofeedback device, for example, which we use to produce information for our own medical needs, or our own entertainment or self-understanding, and not for government to use in the manner of its choice.

FOURTH AMENDMENT COVERAGE – AND PERFECTLY EFFICIENT SEARCH TECHNOLOGY

There is another possible argument that some neuroimaging data will fall outside Fourth Amendment coverage. Recall, again, Justice Stevens's argument in Kyllo that police should be able to freely gather evidence even about what happens inside a home when the methods they use reveal only something dangerous or illegal – such as "the odor of deadly bacteria or chemicals for making a new type of high explosive" (Kyllo v. United States 2000, 48).

This is a search doctrine that Supreme Court had previously made the centerpiece of two cases. United States v. Place dealt with the question of whether the government engaged in a Fourth Amendment search when using a dog – trained to alert when it smelled illegal narcotics – to sniff the air around the luggage of a traveler whom the Drugs Enforcement Administration (DEA) had come to suspect might be carrying drugs. The Court held that the DEA agent's detention of the traveler's luggage was a Fourth Amendment "seizure," and that the length of its detention was unreasonable under the circumstances. The Court, however, also addressed, in the course of its decision, whether the government engages in a Fourth Amendment search when it uses a trained dog to sniff the package for drugs. The Court decided it did not: The dog sniff, said the Court, "does not require opening the luggage" and "does not expose noncontraband items that otherwise would remain hidden from public view, as does, for example, an officer's rummaging through the contents of the luggage." Indeed, said the Court, the dog sniff reveals absolutely nothing except "the presence or absence of narcotics, a contraband item" (United States v. Place 1983). This intrusion was so limited, said the Court that it did not even count as a Fourth Amendment search.

A year later, the Court applied the same analysis to a chemical test a government agent conducted on white powder that Federal Express employees had noticed in a damaged package and called the DEA to report. The DEA agent "removed a trace of the white powder" and administered a field test "on the spot" which "identified the substance as cocaine" (United States v. Jacobsen 1984, 112). As it had in Place, the Court found that the chemical test was not a Fourth Amendment search: "A chemical test that merely discloses whether or not a particular substance is cocaine," it said "does not compromise any legitimate interest in privacy" (United States v. Jacobsen 1984, 123). In fact, said the Court, "even if the results are negative – merely disclosing that the substance is something other than cocaine – such a result reveals nothing of special interest. Congress has decided – and there is no question about its power to do so – to treat the interest in 'privately' possessing cocaine as illegitimate" (United States v. Jacobsen 1984, 123).

In short, each of these cases were real-life versions of Loewy's hypothetical "divining rod," (discussed in Chapter 2) which reveals only the presence of criminal activity – without simultaneously revealing any innocent activity or data (Loewy 1983, 1247). Indeed, Loewy even suggested the use of "a marijuana-sniffing dog sniff" was one of the real life search methods closest to a divining rod – although he also noted that he opposed the "carte blanche use of marijuana-sniffing dogs," and that, to the extent that "the dog is less than perfectly accurate, innocent people run the risk of being searched" (Loewy Arnold 1983, 1247).

With these cases as background, one might ask whether this part of Fourth Amendment doctrine provides another rationale by which certain neuroimaging data might be excluded from Fourth Amendment coverage. If the only thing a certain brain imaging design tests is whether someone's brain activity shows that he is familiar with a murder weapon, might that be a non-search on the ground that no person has a legitimate interest in hiding familiarity with the weapon? Of course, it is possible that, if government infers from a person's P300 response, that he recognized a murder weapon, this inference will be mistaken – or, even if correct, will be detecting a sense of familiarity that doesn't arise from any kind of criminal conduct (just because he finds a weapon familiar-looking does not mean he used it in the crime). But the kinds of dog sniff and chemical tests that the Court held to be non-searches can also show false positives – and this has not dissuaded the Court from excluding them from Fourth Amendment coverage.

Moreover, even when cocaine is genuinely in someone's luggage, this may not be evidence that the person carrying it is engaged in criminal activity: She may have been unaware of its presence, not realizing – for example – that someone else was using her as an unwitting pawn in a delivering the drug. Authorities may still have a powerful interest in detecting the drug. However, in stopping her and searching her suitcase, they are subjecting to government surveillance a space (in her luggage) that she generally has a right to regard as private, and is insulated by the Fourth Amendment from government investigation except when officials have powerful reasons to intrude into it. Why then, one might ask, doesn't the government have an equally good argument that it has a right to use non-intrusive methods that establish only whether someone shows familiarity with key evidence from a crime scene?

There are a few possible answers to this question. One is that a thought or feeling, no matter how closely connected it may seem to criminal activity, cannot itself be a contraband item – in the way that an illegal drug is. When one possesses an illegal drug, one possesses an item one has no right to have. A thought, by contrast, can never be illegal itself. It must always be accompanied by some kind of action to count as criminal. Moreover, tests – like Farwell's brain fingerprinting – do not present only the "probe" stimuli. They also ask the subject to react to other stimuli, and – no matter how mundane these may seem – they can reveal or confirm additional information about the subject's thinking. As a consequence, even if brain imaging is focused only on revealing limited information about a person's familiarity with certain items or images, it is highly unlikely the Court would classify it as so removed from any legitimate privacy interest as to be a non-search.

The fact that it involves probing into a person's brain and mind, and not merely a package or piece of luggage, adds to the case of maintaining Fourth Amendment coverage. And this is strengthened by Michael Adler's net-wide search hypothetical (discussed in Chapter 2). As Adler noted, even a search which – like that in Place or Jacobsen – is perfectly designed to reveal only contraband may seem at odds with a free society when it constantly or frequently brings the government into spaces from which, our society assumes, government must generally be excluded. Adler's example of such an environment is the digital storage space inside of our houses. Even if a particular government search there targets only what is illegal, "a search that eliminates an individual's control over the boundaries to her most private realms would likely be perceived as a threatening exercise of coercive power." What makes this even more damaging is that, as Adler says (Adler 1996, 1112), such a search need not be limited to evidence of actions (such as murder) that are universally and enduringly recognized as wrong. "[V]irtually any socially disfavored act can be criminalized at the discretion of the majority; the individual would then retain no control over whether or not information relevant to such an act would be revealed" (Adler 1996, 1111). Thus, even the most limited probing of our minds for knowledge of illegal activity might be unjustifiably damaging to our mental privacy, and perhaps to others' sense of privacy – even if it reveals only illegal activity. A brain imaging test that reveals only knowledge of who committed a minor act of vandalism, for example, might do more damage to such a sense of privacy than any benefit it brought in terms of crime control. Still, as I will soon argue, where an EEG, fMRI, or other brain imaging test is very limited in what it reveals, this may not make law enforcement use of the technique a non-search, but it will affect the analysis the Court does to determine if such a search is reasonable.

FOURTH AMENDMENT PROTECTION – AND THE IMPORTANCE OF BALANCING

The philosopher Robert Nozick opens his book, *Anarchy, State and Utopia*, with the following characterization of how to think about rights: "Individuals have rights, and there are things no person or group may do to them without violating their rights" (Nozick 1974, xix). American constitutional rights aren't quite so absolute. They apply only against the government, not against private parties. And even the staunchest constitutional protection has exceptions. For example, even though the First Amendment generally protects individuals against ideological censorship, it will permit the government to engage in such censorship on the rare occasion that it has a "compelling interest" in doing so. Still, many constitutional rights are at least approximations of Nozick's ideal. The First Amendment right to freedom of speech and Fifth Amendment privilege against self-incrimination, for example, block government action: When they cover a certain government method, they prevent it (or, the case of the First Amendment, do so most of the time).

By contrast, the Fourth Amendment right against unreasonable search and seizure is somewhat different. While it firmly prohibits unreasonable search and seizure, the uncertainty about what counts as "unreasonable" in specific circumstances makes its protection give way more easily. Indeed, rather than completely shield a particular private realm of human action (such as speech), it is often described by the Court as balancing the individual's interest in privacy against the state's interest in conducting an investigation. "The essential purpose of the proscriptions in the Fourth Amendment," it has said, is to "impose a standard of 'reasonableness' upon the exercise of discretion by government officials," and "the permissibility of a particular law enforcement practice is judged by balancing its intrusion on the individual's Fourth Amendment interests against its promotion of legitimate governmental interests" (Delaware v. Prouse 1979, 654). In the default case, as noted earlier, the warrant requirement constrains how this balancing takes place: Even when law enforcement has very strong interests in searching a home, this typically does not justify police entering on their own discretion. They rather have to make the case to a neutral magistrate that there are "legitimate government interests" in searching the place they have particularly specified, and that they have probable cause to search. Government must, in other words, meet some "objective standard" of reasonableness – such as probable cause or some other level of individualized suspicion – rather than simply argue that its interests outweigh that of the individual (Delaware v. Prouse 1979, 654).

But in many cases, the Court has identified exceptions to the warrant requirement. Sometimes this is because it is impractical to insist on a warrant, or on a showing of probable cause. This is true, for example, where an officer might have to enter a house (or other private area) immediately in "hot pursuit" of a felon, or to deal with exigent circumstances threatening someone's safety, or possibly the destruction of evidence. Those threats – to safety and integrity of evidence – might also arise when police arrest a suspect, who could be carrying arms and has strong incentive to destroy evidence before it can be used against him. In other cases, it is impractical to expect police to know, beforehand, where to find the dangers they are looking for – so it is not only a warrant, but even individualized suspicion, that is impractical. At an airport, for example, the danger feared is extraordinary and is very difficult to pinpoint ahead of time. The same is true, the Court has said, when police try to find drunk drivers or schools try to find students using drugs during school activities. Moreover, a warrant is not only impractical in such environments, it is less necessary because the expectations of privacy people have in the heavily regulated confines of airports or schools, for example, are much lower – and the regulation they face is not first and foremost designed to thwart and prosecute crime, but to advance a "special need" different from ordinary crime control (like assuring school – or transportation – safety).

With the warrant requirement out of the picture, the Court is often thrown back in these cases on fundamental Fourth Amendment purposes – which, on the view it has often taken, require it to carefully weigh the investigation method's "intrusion on the individual's Fourth Amendment interests against its promotion of legitimate governmental interests."

FOURTH AMENDMENT PROTECTION – WARRANTLESS SEARCHES, AND PROBLEMS WITH BALANCING

Use of such a balancing approach, however, is problematic for administrative and special-needs searches, and many other types of warrantless searches. In the first place, under a balancing regime, when security or safety concerns become powerful enough, as they are in the wake of terrorism attacks, Fourth Amendment interests in privacy or autonomy become more and more likely to find themselves in an unwinnable contest. Not surprisingly, when balancing has been applied in the context of special-needs and administrative search cases, for example, the government almost always wins. Moreover, the balancing has also frequently favored police when they conduct a search incident to arrest and, for example, administer a breathalyzer to obtain evidence of drunk driving. Recall Kerr's explanation of how the Fourth Amendment divides law enforcement investigation into "less invasive steps" free of constitutional oversight, and "more invasive steps," that count as "searches" and are subject to warrant requirements or other constitutional limits. We might say that special needs and administrative searches represent a kind of hybrid of the two sides of this division: They are searches, because they implicate realms of privacy covered by the Fourth Amendment, but – thanks to the dangers government needs to combat, the lowered privacy interests at stake, or both – officials often get a level of investigative leeway approaching that which they have when they conduct less invasive steps outside of the Fourth Amendment's coverage.

It is this thumb on the scales in favor of warrantless government searching that has caused some scholars thinking ahead – to Fourth Amendment applications to neuroimaging – to fret, especially since many of the searches the Supreme Court and other courts permit under such balancing regimes involve searches into the body. For example, in Skinner v. Railyway Labor Executives Association (1989), the Court respectively allowed government to administer random drug tests – through blood tests, breathalyzer tests, and urine tests – in the railway industry. In Vernonia School District v. Acton (1995) and Board of Ed. of Independent School Dist. No. 92 of Pottawatomie Cty. v. Earls (2002), it permitted random drug tests of public school students – tests that required collection of urine samples. In airports, as I have said, courts have permitted use of magnetometers, and more recently, powerful millimeter scanning of individuals' persons, to assure passenger safety.

If, as noted earlier, the clearest basis for the Fourth Amendment to cover neuroimaging is that it is, like a search into urine or breath, and intrusion into our bodies, or that it is like an X-ray, then – the template of the warrantless search cases above indicates we might be subjected to compelled neuroimaging in a number of circumstances where police wouldn't need a warrant use this technology. It is worth looking more closely at two of these circumstances: (1) special-needs and administrative searches, and (2) searches incident to arrest and other cases where exigency might justify a search that would otherwise require a warrant.

For example, to illustrate why neuroimaging might conceivably, like other searches of our bodies, be something police can do without a warrant, Pustilnik cites Schmerber v. California, where the Court found it was reasonable for police to compel the drawing of Schmerber's blood "incident to [his] arrest." Given that "the percentage of alcohol in the blood begins to diminish shortly after drinking stops, as the body functions to eliminate it from the system," there may have been no time for the officer to obtain a warrant – and the blood draw was not unreasonably intrusive to Schmerber, or damaging of his dignity (Schmerber v. California 1966, 770–771). Lower courts have since often emphasized that the possibility of losing the evidence in that case presented "exigent circumstances." In wondering whether Fourth Amendment protections might be too weak to protect against warrantless neuroimaging without modification of the doctrine, Farahany takes note of Skinner v. Railway Labor Executive Association, and the warrantless blood, urine, and breath tests it permitted as part of a "special-needs" search (Farahany 2012b, Searching Secrets, 1263–1264). She also notes that, when it comes to concerns about bodily intrusion, the court's limits on law enforcement are "certainly not absolute" and the "Court has found such procedures to be reasonable searches so long as the test is routine and minimally physically invasive" (Farahany 2012b, Searching Secrets, 1284). But neuroimaging tests, as she notes, generally "would not be physically invasive" (Farahany 2012b, Searching Secrets, 1288).

Both Pustilnik and Farahany are understandably concerned about the prospect that – in circumstances like these, where a warrant is not required, and the Court turns to balancing – neuroimaging will be treated in the same way as non-intrusive testing of one's body. Pustilnik, for example, worries about the possibility that courts might treat "thoughts" as equivalent to "fingernail clippings" or other "physical samples" (Pustilnik 2013, 131). In part, the problem stems from a story about Fourth Amendment coverage that relies entirely on brain activity taking place in the interior of the body (like blood flow, air intake, or formation of urine). As she points out, "[l]ooking at our brain emanations as physical samples apart from their informational content and apart from the extent to which mental privacy allows us to constitute our identities" provide us with only a deeply "impoverished" account of the privacy interests at stake (Pustilnik 2013, 131). We seem to be missing a key part of the story when we treat our brain activity as being private only in the same sense as our blood, breath, or urine are private. Farahany similarly argues that courts ignore a crucial privacy interest when they give brain activity Fourth Amendment protection only to the extent that it is (like other physiological processes) secluded within the body (Farahany 2012b, Searching Secrets, 1288–1289). The Fourth Amendment, she argues, should not protect brain activity simply because of the seclusion that characterizes it, but because of the secrecy of the thoughts and memories that could be compromised if government could freely observe such activity. These concerns echo those of scholars who have worried that existing Fourth Amendment law (including both third party doctrine and some warrantless search cases) provide government with too much access to external records of our thinking – and particularly those we leave in digital form (Solove 2007, 112).

What then is to be done if government wishes to conduct neuroimaging at airports or at entrances to a federal building? This example is not entirely fanciful. Government has already used behavioral profiling in airports to try to identify dangerous intentions (and not simply the dangerous items it looks for in electronic airport screening). Justin Florence and Robert Friedman describe the Department of Homeland Security efforts to use behavioral profiling (Florence & Friedman 2010, 425–430). Christopher Rogers writes of efforts to explore supplementing this with biometric "Future Attribute Screening Technology (or FAST), which can remotely read a person's vital signs and then predict whether that person has the indicators of 'malintent,' the intention to commit a crime" (Rogers 2014, 339–340). Farahany also takes note of an Israeli company, WeCU, marketing a device that presents passengers with subliminal stimuli at airport check-in lines – in order to generate emotional responses because that "emotional response is highly predictive of a passenger's potential security threat" (Farahany 2012a, Incriminating Thoughts, 376). As others have suggested, it is not far-fetched to think government may use technology like this to engage in behavioral profiling by looking at brain activity (Federspiel 2008, 889, 890–91), (Holley 2009, 13), (Moreno 2009, 717–719). Neuroimaging could conceivably allow the government to gather additional information of the kind these profiling systems and technologies are meant to enable.

FOURTH AMENDMENT PROTECTION – TWO POSSIBLE ALTERNATIVES TO THE STATUS QUO

One possible response to such concerns is to accept the basic Fourth Amendment framework as it is – that is, one where courts balance the intrusion into privacy against the government interests advanced, but then insist that in doing so, courts must accurately assess the privacy intrusion that neuroimaging causes, looking not just at how physically intrusive it is, but also at how mentally intrusive it might be. Moreover, there are at least some hints in the Court's past Fourth Amendment cases that it would be responsive to such arguments. For example, the Court has emphasized that whereas urine analysis does not require the intrusion into the body that occurs in a blood test, it "can reveal a host of private medical facts about an employee" (Skinner 1989, 616). Similarly, it might acknowledge that no matter how minimal the physical intrusion is in neuroimaging (and possibly even less intrusive in neuroimaging of the future), analysis of recorded brain activity can reveal numerous facts about a person's thought. Moreover, in its search incident-to-arrest line of cases, the Court has recently been willing to change the rules to protect private information (in a cellphone). (I say more about this later) Perhaps it will similarly change rules for special-needs and other warrantless searches if and when those searches make use of neuroimaging technology.

However, it is worth considering two other possible responses to such concerns. First courts could hold that while many different investigative techniques may be fair game for authorities in these kinds of searches-incident-to-arrest, special-needs searches, and other warrantless searches, neuroimaging simply isn't. In other words, courts could adopt an across-the-board rule that any kind of search that may intrude into mental privacy requires at least a warrant and probable cause (if not more). And so, schools and workplaces, for example, simply may not use neurotechnology to examine people about whom they have no suspicion. The idea behind this approach is that when a search method is so intrusive that it threatens mental privacy, it needs at least the kind of protection provided by an individualized determination of whether a warrant is justified in a particular case.

In fact, this is precisely what the Court has done in some of its recent search incident to arrest cases. As a general matter, the Court has for decades allowed police to search individuals whom they arrest – immediately and without a warrant – for two reasons: (1) to check the arrestee for weapons to see if he presents a threat to the officer and (2) to prevent the arrestee from destroying evidence. When either of these were a possibility, an officer could search any item or area within the arrestee's control (Chimel v. Califronia, 1969). Moreover, the Court said in United States v. Robinson that officers could search the arrestee's own person even *without* showing that there was a danger to the officer's safety, or of destruction of evidence. In other words, it adopted a categorical rule that searching an arrestee's person without a warrant was permissible (United States v. Robinson, 1973). Thus, even where officers had no valid reason to worry that a cigarette package they had found in Robinson's coat pocket presented a safety risk or had to be opened to preserve evidence related to Robinson's alleged crime, they could warrantlessly open this package (as they did, discovering drugs hidden within it). In 2014, however, the Court modified the rule in Robinson: Instead of assuming that police could *always* search the person of an arrestee, the Court said that – confronted with the question of whether officers could search SmartPhones obtained from the person of an arrestee – judges should not simply apply Robinson's rule "mechanically" but should instead go back to Fourth Amendment first principles. They should determine – if a warrant is necessary by "assessing, on the one hand, the degree to which [a search] intrudes upon an individual's privacy and, on the other, the degree to which it is needed for the promotion of legitimate governmental interests"

(Riley v. California, 2014). In the case of a cell phone, the Court said, a warrantless search intruded too deeply into individual privacy interests. "Cell phones," said the Court, "place vast quantities of personal information literally in the hands of individuals. A search of the information on a cell phone bears little resemblance to the type of brief physical search considered in Robinson" (Riley v. California, 2014).

The Court applied the balancing test again to analyze two related search-incident-to-arrest questions in the 2016 case of Birchfield v. North Dakota: the question of whether police need a warrant to search a drunk driving arrestee (1) with a blood test and (2) with a breathalyzer test (Birchfield v. North Dakota 2016). Applying the same balancing test described in Riley v. California and numerous other cases, the Court reached a split decision: blood tests of arrestees require warrants, breathalyzer tests do not. The reason for the different outcomes largely lay in the privacy intrusion inherent in each test: A blood test requires breaking the skin and removing blood. It also potentially reveals any information a chemical test might reveal in the blood. A breathalyzer test uses air that would be released by the body in any case. Moreover, said the Court the breathalyzer test is designed to provide only information about alcohol content, and nothing else: authorities do not retain the breath sample in such a way that they could extract other information from it (Birchfield v. North Dakota 2016).

The result of such analyses is that certain kinds of tests police wish to conduct incident to an arrest – a search of a cellphone, for example, or a blood test, will now typically require a warrant. They are removed from the realm of permissible objects of a warrantless search unless some other exception applies (for example, because there is demonstrable exigent circumstance that requires access to a cell phone that would otherwise require a warrant). In other words, while searches of persons incident to arrest have been treated as akin to "less invasive" steps where police have considerable leeway to search vigorously (even without specific reasons for focusing on a particular item), searches of their cell phone data have now been moved back to the "more invasive" category – in part, because, when it comes to the intangible information in SmartPhones, there is less immediate threat to law enforcement interests from such information, and there is also a significantly stronger privacy interest at stake (Riley, 2014, 2485). Neuroimaging could likewise typically require a warrant or other showing of justification even in many cases where other kinds of police investigations can take place without one (such as in an airport) for similar reasons: Because mental states are not, by themselves, danger-ous in the way that explosives or other weapons are, and because there we have far greater privacy in our thoughts than in our luggage. (For the reasons noted in Chapter 2, the weight of this privacy interest may depend on the nature of the neuroimaging technique used and the specific infer-ences about thought it makes possible.)

FOURTH AMENDMENT PROTECTION AND FIRST AMENDMENT PROCEDURES – A THIRD ALTERNATIVE TO THE STATUS QUO

However, there is still another possible response – and I will argue that it is an advisable one, not instead of an approach that emphasizes the impor-tance of mental privacy, but in addition to such an approach. In short, the Court's framework for permitting special needs and certain other warrant-less searches will be better – and better-equipped to deal with technologies such as neuroimaging – if it moves away from unpredictable balancing of privacy and security interests to which it is hard to assign a definite weight (and thus, hard to balance in a principled way).

Such a shift would have benefits for Fourth Amendment law that go beyond its application to neuroimaging. Blanket searches in some respects resemble the kind of "general warrant" the Fourth Amendment was designed to forbid: The Fourth Amendment's text requires officials specify the place to be searched, and have probable cause to search it. It does not permit them to do what officers of the English crown used to do with general warrants, which is to explore numerous papers, places, or other possible sites of criminal evidence – even without any strong reason to focus on them. But in special-needs and administrative searches, this is precisely what the government does. Christopher Slobogin describes searches such as "special needs" and administrative searches as examples of what he calls "panvasive" searches, by which he means "modern govern-ment's efforts at keeping tabs on the citizenry [that] routinely and randomly reach across huge numbers of people, most of whom are innocent of any wrongdoing." And he notes there is a tension between the Court's permis-sive attitude toward such searches and the Fourth Amendment's widely known inconsistency with general searches. As Slobogin recognizes, this presents a problem – and requires some constitutional constraint that can make up for the absence of the warrant requirement, and of particularized suspicion (Slobogin 2014, 1722–1723).

The Court has hinted at such requirements. In the context of administrative searches, of regulated businesses, it has done more than hint. It has stressed the existence of what it calls "a constitutionally adequate substitute for a warrant." Rather than leaving the inspection and penalizing of regulated business subject "to the unchecked discretion of Government officers," there is a federal statute governing the industry which "establishes a predictable and guided federal regulatory presence" (Dewey v. Donovan, 1981, 604). In the context of special-needs searches, like random drug testing, it has been less clear. But, as I have argued in previous scholarship, (Blitz 2004, 1457–1478), the Court often emphasizes features of the blanket search that it argues, in some sense, make up for the absence of warrant.

For example, the Court typically takes note of whether the government's blanket search is characterized by certain features that effectively minimize the intrusion it makes into an individual's privacy and dignity. They often note, for example, that officials have little control over how a search of this kind will be conducted or what kind of information it will turn up: In drug tests, for example, there is a very specific protocol from which officials cannot easily deviate. Also, such tests will often reveal only the presence or absence of a certain drug in the blood, breath, or urine. And the tests' administration is confined to an environment that, as I have noted above, is heavily regulated. A train worker is tested for alcohol under such a regime not when at home or on vacation, but when she is operating trains – conducting a task where it is in the public interest to assure she is able to concentrate.

These hints suggest another possible response to the possibility that neuroimaging techniques might find their way into special needs, administrative or other warrantless searches. The key problem with balancing, as I noted earlier, is that where security interests are deemed to be very high, as they have been in cases where drug use threatens safety, or where violent attacks or accidents are possible in car or air travel, then it is likely they will be deemed by courts to justify even highly invasive searches – regardless of whether the government could do better in protecting privacy. As a consequence, even where the government can do without neuroimaging, courts may let the government use it anyway if it determines that the balance of interests is heavily in the government's favor (as it typically is in special-needs cases).

An alternative regime modeled on First Amendment law, however, would not give the government such an option. Free speech rights, as I noted, are closer in nature to Robert Nozick's barrier against government interference that Fourth Amendment rights typically are. And the specific material that barrier is made of – in First Amendment and other constitutional contexts – is "heightened scrutiny" on the part of the judiciary towards any kind of government action that threatens the right in question (Sorrell v. IMS Health, 2011, 2664). More specifically, in order to justify a speech restriction, government has to show (1) it has a very important goal – one which can justify a measure as worrisome and unusual as placing limits on speech – and (2) that something as foreign to a free society as speech restriction is, in this case, necessary to achieve that goal, but that they are taking measures to restrict speech as little as possible (or at least, not significantly more than necessary) to further their goal. Sometimes, this form of heightened scrutiny takes on a form ("strict scrutiny") which is almost impossible for the government to satisfy: When the government wishes to suppress speech on the basis of the message it carries, for example, the Court will almost reject such ideological censorship – allowing it only when the government has an extraordinarily important interest (or the kind a court refers to as a "compelling interest"), and can achieve that interest in no other way (United States v. Playboy Entertainment, 2000, 812–813, 816, 817). Even then, the court will require the government to use the "least [speech] restrictive" measure available to it – so that it avoids doing any more damage than necessary to First Amendment interests. In other First Amendment cases, the court lowers the bar a little bit and applies only "intermediate scrutiny." In these cases government has to cite only a "substantial interest" which need not be as rare or extraordinary as one it would need in other cases where strict scrutiny applies and its restriction, while it need not be a perfect fit with the government's ends, must avoid burdening "substantially more" speech than necessary to achieve these ends (Ward v. Rock Against Racism, 1989, 794–796, 799).

The Fourth Amendment typically does not operate like this: Police can obtain a warrant to investigate any crime. They do not need to show the judge that the particular crime they are investigating is a particularly grave one that gives them a compelling or significant interest in obtaining a warrant. Nor do they need to show, in each case, that they will minimize the intrusion. To be sure, they do have to specify a particular place they will search – so that they will not, as under a general warrant, have access to all of a person's papers, possessions, or private space. But in many cases when a warrantless search is permissible, even this limit is gone, or substantially weakened.

However, when extraordinary privacy interests are at stake in such a balancing situation, the better response is for a court to shift from vague balancing to a more constrained inquiry that requires government not merely to show that its interest are strong, but also that it is minimizing the harm done to the interests at stake. This is especially appropriate when the interests are not only Fourth Amendment interests, but simultaneously First Amendment interests. Courts should not be indifferent to situations where government does those interests substantially more damage than is necessary to achieve its objectives. Indeed, as noted above, courts have already occasionally emphasized minimization where they see it in a special needs case: They have noted, for example, that drug tests leave little room for government discretion (Vernonia School District v. Acton 1995, 658, National Treasury Employees Union v. Von Raab 1989, 667) and – harkening back to the decisions in Place and Jacobsen on dog sniffs and chemical drug tests – reveal nothing more than the presence of illegal drugs in a person's body (Skinner 1989, 672 n.2).

More specifically, government effectively should have to meet intermediate scrutiny in order to assure that is not doing unnecessary and excessive damage to Fourth and First Amendment interests. Under this regime, government should have to show, for example, that the problems it is addressing with neuroimaging justify such strong law enforcement medicine. The need to detect and thwart terrorism or other violent crime might qualify as such an interest. So too might the need to detect financial crime (such as identity theft) that causes tremendous disruption to society and is difficult to detect with normal law enforcement means. But even so, government should not be permitted to use such methods in special-needs searches where there are other methods of available that are less threatening to individuals' privacy and freedom of thought.

Consequently, even where it can show that neuroimaging is necessary in a special-needs search, for example, it should also have to show that it has built into its general search regime privacy protections that help assure that the invasion created by the search is not far greater than necessary. Solove recommends that such heightened scrutiny be applied to government attempts to obtain and gain external records of our thoughts – in diaries or Web searches, for example (Solove 2007, 151–176). There is an equally strong case for applying it to government access to our unexpressed thoughts. For example, courts should insist if government needs access to mental states, that (where possible) it employ technology that probes only those aspects of a person's thinking it needs knowledge of to serve the significant government interest in question. A variant of this approach to modifying search and seizure doctrine might also be applied when information is collected through use of grand jury subpoenas rather than in a police search. Pardo and Pustilnik have noted that Fourth Amendment protections are considerably weakened in cases involving such subpoenas. In Dionisio v. United States, a case discussed by Pardo (Pardo, 327, 2006), the Supreme Court observed that the Fourth Amendment provided protection against a "grand jury subpoena duces tecum" – that is a subpoena to produce certain tangible items – "too sweeping in its terms 'to be regarded as reasonable.'" But it went on to find that the defendant did not have the kind of privacy interest that would justify raising a Fourth Amendment shield to prevent subpoena of the voice exemplar demanded of Dionisio (which law enforcement wished to compare with recordings in evidence): The "physical characteristics of a person's voice, its tone and manner, as opposed to the content of a specific conversation, are constantly exposed to the public," said the Court, and therefore unprotected (Dionisio, 1973, 11, 14). Pustilnik likewise discusses cases where non-intrusive gathering of information about biological activity was permitted in a grand jury investigation (Pustilnik, 2013, 132–134). As in warrantless searches, one could argue that a framework for assessing the permissibility of such a subpoena should accord sufficient weight to the full privacy interests at stake in neuroimaging (and not just their physical invasiveness).

FOURTH AMENDMENT PROTECTION – BEYOND THE WARRANT REQUIREMENT

Having argued that this kind of First Amendment regime should replace the balancing the Court does in some situations where warrantless searches are normally permissible (especially when they are blanket searches), I also argue that such a First Amendment heightened scrutiny structure should, in some cases, be layered over the warrant requirement – even where that requirement does apply – and that there is a particularly compelling case for doing so when the government's search technique involves neuroimaging.

If imaging of someone is a "search," then – unless one of the recognized exceptions apply – police will only be able to use such a technique to gather information from an individual if they first obtain a warrant based upon probable cause. But is that sufficient protection for compelled brain imaging, even in its current limited form? Would it be sufficient protection for brain imaging of the future, that might occur without the subject realizing it is occurring, and may, perhaps, be able to the gather more detailed information about the person's psychological characteristics or about mental events? The warrant and probable cause requirements certainly provide a safeguard against arbitrary or groundless searches of a person. They require police officials to build a case – before using a certain kind of investigatory method – that shows that that there is a "fair probability" that they will find evidence of criminal activity in the place to be searched.

But one might argue that when police wish to search in certain places, or with certain very intrusive techniques, even more should be required. Michael Pardo and Dennis Patterson note that the Court has imposed "probable cause plus" requirements for a warrant, but that these apply under current Fourth Amendment law only when "a search or seizure poses physical risk to a defendant," not on the basis of the privacy of the information sought (Pardo and Patterson, 213, 154, n. 37).

There is, however, another circumstance in which courts (especially lower courts) demand more of police than a warrant: when police engage in wiretapping or in video-surveillance of the inside of a home or other private space. In New York v Berger, the 1967 case in which the Court first extended the Fourth Amendment to cover wiretapping, the Court insisted that police seeking a warrant need to specify more than "probable cause" and "the place to be searched." Because electronic eavesdropping "involve[s] a privacy violation that is broad in scope," said the Court, it imposes "a heavier responsibility" on a court "in its supervision of the fairness of procedures." Apart from simply describing a target in describing the place to be searched, police had to – in meeting this particularity requirement – describe the "type of conversation sought with particularity, thus indicating the specific objective of the Government." They also had to limit their intrusion to "one limited intrusion, rather than a series or a continuous surveillance." Such safeguards, said the Court, insured that the "danger of an unlawful search and seizure was minimized" (Berger 1967, 57–58).

In the wake of Berger, Congress used the Court's discussion in that case as a template for imposing specific statutory requirements on federal wiretapping. Apart from requiring a judge to assure that (1) that the warrant must contain "a particular description of the type of communication sought to be intercepted, and a statement of the particular offense to which it relates," Congress also imposed other requirements, including that (2) "normal investigative procedures have been tried and have failed or reasonably appear to be unlikely to succeed if tried or to be too dangerous," (3) that the time during which the surveillance of a conversation is to take place is not longer than is necessary to achieve the objective of the authorization, nor in any event longer than thirty days" and (4) that a wiretap interception of a telephone conversation "be conducted in such a way as to minimize the interception of communications" that are not related to criminal activity subject to investigation (18 U.S.C. Sections 2518(3)-(5)).

These requirements were imposed by Congress, in its wiretap act, not by the Constitution in the Fourth Amendment. But beginning in 1984, a number of federal courts imported these requirements into Fourth Amendment law when the government sought to use surreptitious video-surveillance in a private environment. In United States v. Torres, the Seventh Circuit Court of Appeals found it "unarguable that television surveillance is exceedingly intrusive, especially in combination with audio surveillance, and inherently indiscriminate, and that it could be grossly abused – to eliminate personal privacy as understood in modern Western nations." It thus imposed the same requirements on television surveillance that Congress had imposed on wiretapping in above-cited statutory provisions, finding in the constitutional context, and with respect to such an intrusive surveillance method, they were necessary to "implement the constitutional requirement of particularity" (Torres 1984, 884). Since then, a series of federal courts across the United States applied the same or similar requirements in other video surveillance cases.

Should courts impose similar requirements on technology that is used to gather physiological data from the brain rather than from wiretaps or video surveillance? Susan Freiwald argues the justification for such "heightened procedural hurdles" beyond the ordinary warrant requirement depends on the four features of surreptitious video surveillance and wiretapping: that it is (1) "hidden," (2) "intrusive," (3) "indiscriminate," and (4) "continuous." Hidden surveillance justifies a higher hurdle because "the target is less able to hold government investigators accountable, and therefore needs the court to protect his interests." Intrusive methods do so because they "bring law enforcement further into our private lives, and therefore require judicial intervention to ensure that government makes such intrusions only after satisfying a high level of need." Surveillance is indiscriminate when it "obtains information beyond that which is justified, and thus requires court oversight to ensure unjustified surveillance is minimized." And "continuous surveillance is more likely to be intrusive and indiscriminate because it acquires more information over a longer period of time" (Friewald 2007, 10–11).

The feature of this quartet of characteristics that has been more heavily emphasized by courts applying the Fourth Amendment, and also by many other scholars, is the indiscriminate nature of some searches. This worry about indiscriminate searches flows naturally from the particularity requirement in the Fourth Amendment. In order to obtain a warrant, as I have said, police need to do more than simply show that they have probable cause to believe that their investigation of certain activity will produce evidence of a crime. The Constitution's text itself makes it clear that they must also "particularly describ[e] the place to be searched, and the person of thing to be seized." This "particularity" requirement is intended to forbid – in US enforcement of the law – that practice of using "general warrants," which, as the US Supreme Court said, "allowed royal officials" in eighteenth-century England and the English colonies, "to search and seize whatever and whomever they pleased while investigating crimes or affronts to the Crown" (Stanford v. Texas 1965, 472).

Armed with a general warrant, officials could enter any house or other private space they have a hunch might aid their investigation, and then search whatever they wished to in a person's home. They could, in other words, engage in what the Supreme Court described as "general, exploratory rummaging in a person's belongings" (Coolidge v. New Hampshire 1971, 443). The particularity requirement shields individuals in a free society from this kind of unconstrained invasion into their personal spaces and possessions: It not only forces police to justify their search of such spaces or possession to a magistrate – and obtain a warrant from that magistrate. It also requires that the warrant itself be limited in scope, so that the search focuses only on that part of a person's space or possession that police reasonably believe is likely to contain evidence of a particular crime.

Yet courts and scholars have argued that some emerging technologies make it hard to prevent "general exploratory rummaging" of people's information. This was, in large part, why courts have insisted on minimization requirements for wiretapping and video surveillance: Such electronic surveillance will likely capture not only conversation, or images, concerning criminal activities, but all other words that are exchanged in a phone call or in front of a hidden video camera.

Thus, David Gray and Danielle Citron argue that the Fourth Amendment should apply not only to protect us in traditionally private areas, but also against technologies that subject us to "broad programs of indiscriminate surveillance" (Gray and Citron 2013, 73). In recent years, some magistrate judges have also imposed specific limitations on computer searches on the grounds that such searches will otherwise give government access to numerous files that have little to do with the crime it is targeting. In United States v. Comprehensive Drug Testing, for example – a case in which federal agents searched computer files for information about specific major league baseball players' steroid use – the magistrate issued a warrant for a search of computer files, but imposed "significant restrictions on how the seized data were to be handled. These procedures were designed to ensure that data beyond the scope of the warrant would not fall into the hands of the investigating agents" (United States v. Comprehensive Drug Testing 2010, 1166). As Orin Kerr writes, it has become increasingly common for magistrate judges to impose ex ante restrictions in computer searches (Kerr 2010, 1243, 1248–1271). As Paul Ohm explains, the logic of such ex ante restrictions is to prevent something akin to a general warrant from being issued under the guise of one that meets the particularity requirement: "Many government practices have been compared to general warrants, but almost none are close to being as invasive as a months-long trawl through a person's personal computer" (Ohm 2011, 10).

To the extent indiscriminate searches remain the primary candidate for additional constraints on the warrant requirement, it is unlikely that current neuroimaging techniques qualify. Where officials use a test that only reveals how individuals' brain activities respond to specific stimuli (or when making specific statements) related to a particular crime, they will not be as likely to come across entirely unrelated activity, as they will when they are listening in on a phone conversation, videotaping unanticipated activity in a living room, or browsing through hundreds or thousands of computer files stored on a computer they have seized. If, for example, the government asks a researcher trained in neuroimaging techniques to tell them whether or not a tested subject's brain responds to an image of a murder weapon, and does so in a way that past studies have correlated with showing familiarity with that weapon, agents will not easily be able to learn much else about the suspect's thoughts or mental states. Consequently, it is likely that a warrant is all that is required here.

Moreover, existing types of neuroimaging technology also are non-candidates for heightened procedures under two of Freiwald's other criteria. They are not hidden: If, in future years, development of infrared imaging or other technology allows a government to surreptitiously and

remotely gather data about brain activity from an unsuspecting individual, then this would of course significantly strengthen the case for
demanding that the government should need more than a warrant –
unless, perhaps, even such surreptitious probing is constrained to reveal
nothing more than a link to knowledge of a crime under investigation.
Similarly, if and when investigators can gather data through brain-
computer interfaces by "camoufloug[ing]" the interrogation of a user,
in the manner described by Martinovic and his colleagues, this too
might be a technique that should be subject to heightened procedural
requirement – for example, the requirement that it be used only when
other means of obtaining the information needed by police have failed,
leaving use of a brain-computer interface as a last resort (Martinovic
et al. 2012). But the neuroimaging methods of the present do not raise
this problem. Likewise, neuroimaging is not continuous, as is an hours-
long wiretap recording or hidden video-surveillance recording. While it
may take a long time to obtain the required images, this is sometimes
because many fMRI scans are needed to gather reliable information
about a particular brain response – not because those using the fMRI
are gathering significant information about person's on-going daily
routine (whether in her communications, or home activities, or choices
about where to drive) over an extended period. If and when future
versions of neuroimaging allow researchers to construct something
more like a continuous video of a person's internal life from second to
second, then this factor might weigh in favor of heightened requirements. The same might be true of surreptitious and portable methods of
neurorecording that allow individuals to take multiple measurements of
an individual's activity over a long period of time, as that individual
performed normal routines (rather than specific tasks dictated by a
supervising researcher or tester).

   There is, however, one of Freiwald's factors that does weigh in favor
of heightened requirements for obtaining a warrant and that is intrusiveness. As noted above, Freiwald suggests that high intrusiveness
weighs in favor of additional safeguards because it "bring[s] the law
enforcement further into our private lives, and therefore require[s]
judicial intervention to ensure that government makes such intrusions
only after satisfying a high level of need" (Freiwald 2007, 10). There is
a strong case to be made that even relatively brief and fragmentary
glimpses of how the brain behaves as it generates our thoughts entail a
level of intrusiveness than justify strong safeguards – and requir[ing]

that government first show that it has "a high level of need" before it can observe such brain activity. As I noted in the introduction, our thoughts and feelings have long seemed to made entirely secure against external observation by the natural order of things. One might argue that this should remain the case – even in a world where neuroimaging can reveal solid clues as to unexpressed thoughts – unless police can demonstrate an extraordinary need for access to this realm of our experience.

And the fact that neuroimaging implicates not only Fourth Amendment – but also First Amendment concerns   provides another argument for treating even simple neuroimaging as deeply intrusive – and intrusive enough to impose additional requirement beyond a simple warrant requirement. If, as the Court has said, the First Amendment protects not only freedom of speech, but also "the freedom of thought" underlying that speech – then police intrusion into the realm of our thoughts is a threat to First Amendment interests.

INTELLECTUAL PROPERTY AND INTELLECTUAL PRIVACY

There are alternative proposals for refashioning Fourth Amendment law to provide greater protection for the mental privacy that may be vulnerable in neuroimaging. One such proposal comes from Nita Farahany. She argues that the courts will better appreciate the privacy interests at stake in neuroimaging if they view at least some of our mental content as similar to intellectual property. "Authors," she argues, "can properly claim a 'secrecy interest' in 'their' writings and effects," and this will sometimes give them a right to prevent an examination of their work even when it is accessible (Farahany 2012b, Searching Secrets, 1243). Moreover, she argues, this authorship right should apply not only to what we write on paper or other media in the external world, but also to memories we refrain from writing about: "Two individuals looking across a courtyard, for example, will focus on different aspects of the scene before them. The memories they encode of that moment will be personally created expressions of their own experiences" (Farahany 2012b, Searching Secrets, 1294). Such an authorship interest, where it exists, adds to the privacy interests in brain activity that derive solely from bodily seclusion. With authorship as an additional principle of Fourth Amendment protection – apart from the privacy we derive from concealing ourselves, our papers, or our effects – Farahany then tests, against this standard, each of the four categories in the spectrum of activity she proposed to replace the testimonial-physical distinction in the Fifth Amendment context: identifying information, automatic processes, memorialized content, and utterances. Her conclusion is that an individual:

> has the strongest claim of authorship in uttered and memorialized evidence and the weakest claim of authorship in automatic or identifying evidence. Because memorialized and potentially recorded utterances are the proper subject of copyright protection, a court must balance the intrusion upon both the seclusion and the secrecy of the individual against the governmental interest in the evidence sought to decide if an unreasonable search or seizure has occurred. (Farahany 2012b, Searching Secrets, 1275)

As was true in her Fifth Amendment analysis of these categories, Farahany finds the Fourth Amendment analysis worrying and counterintuitive in some respects: "If real and intellectual property law are the only sources to which the Court will turn to inform reasonable expectations of privacy in the Fourth Amendment," she notes, then automatically-generated information (including) cognition, may be left without constitutional protection. In some circumstances, this is "the very information that individuals wish to keep the most private" (Farahany 2012b, Searching Secrets, 1308).

There is an answer to this problem and that is that the Supreme Court has not limited privacy interests to real and intellectual property. As noted earlier, even when blood or urine is removed from our body and is no longer our property, the Court recognizes that the "private medical facts" government can draw from it raises distinctive Fourth Amendment privacy concerns about blood and urine testing for drugs. This is not because individuals author such facts, or own them in any other sense. Moreover, we may have strong privacy interests in our thinking processes that are quite different from the intellectual property interests that authors have in their works: When we read someone else's work on an electronic reader or listen to someone else's song, for example, the copyright belongs not to us, but to the owner of the work. That owner is the one who has the right to control what happens to the work and when it can be accessible for others. But government spying on my reading or my choice of what music to listen to would nonetheless be an invasion of my intellectual privacy, even if it is not an invasion of my intellectual property (Cohen 1996, 983–989). One can argue that the memories I create when I read or listen to a work, and the way they are shaped by my distinctive understanding or hearing of it, makes me an author at the very least of my own perceptions. But this move isn't necessary to ground the claim that privacy and freedom of thought should cover those perceptions – which has power not because I author them, but because they are mine (and, by default, generally inaccessible to the rest of the world), whether they consist of identifying, automatic, memorialized, or uttered content. That a piece of intangible information is my intellectual property, or akin to such intellectual property, is something that can certainly weigh in favor of finding that I have a reasonable expectation of privacy in that information. But it is not a necessary condition for such privacy to exist and provide a basis for Fourth Amendment protection.

In fact, among the kind of mental content that many individuals intuitively find to be the most private is mental content that defines features of their personality, even where it consists of "identifying" information. As noted earlier, some neuroimaging could conceivably probe a person's "connectome" (Seung 2012, 4–5) to try to "brainotype" that person (along the same lines probing of genome might aid an observer in genotyping her) (Farah et al. 2010, 126). Such mental-process information seems to be identifying, but it is still deeply private and should, like other mental-content information, receive robust protection against searches unless government can justify its need for access to it. What Farahany says is true of utterances – "when balancing government interests against the fortress of seclusion around the brain, only extraordinary circumstances should justify an intrusion" – should be true in other circumstances that freedom of thought is at stake in the Fourth Amendment context. And, as I have argued, First Amendment heightened scrutiny provides a ready-made model for how to secure such a "fortress" (Farahany 2012b, Searching Secrets, 1308).

CHAPTER 6

Conclusion

Abstract In the conclusion, I argue that when courts apply the Fourth or Fifth Amendment to neuroimaging, they should focus not only on measuring the intrusion that use of this technology creates in a particular instance, but at how use of it might affect individuals' sense of privacy in their mental lives more generally, If privacy of thought is to continue to serve, in Isaiah Berlin's words, as an "inner citadel," then constitutional safeguards should prevent it from being technologically breached not only in cases where the breach would reveal mental content that courts regard as particularly intimate or sensitive, but in all circumstances where the government lacks very powerful reasons for its intrusion.

Keywords Autonomy · Brain · Constitution · Intellectual privacy · Internet surveillance · Isaiah Berlin · Mind · Neuroimaging · Privacy

The image Farahany uses to describe mental privacy – as protected by a "fortress of seclusion around the brain" (Farahany 2012b, 1308) – is an appropriate one, and one that echoes the way other thinkers have described the mind. In his essay, "Two Concepts of Liberty," for example, Isaiah Berlin speaks of how certain thinkers think of their internal life as "an inner citadel" where "there alone" a person can be secure (Berlin 1966, 20).

© The Author(s) 2017
M.J. Blitz, *Searching Minds by Scanning Brains*,
Palgrave Studies in Law, Neuroscience, and Human Behavior,
DOI 10.1007/978-3-319-50004-1_6

Such imagery has found its way into judicial analysis as well: Stoller and Wolpe quote a self-incrimination case that describes the self-incrimination privilege as respecting a "private inner sanctum of individual feeling and thought" (Stoller & Wolpe 2007, 370; Couch v. United States, 1973, 327). In an earlier First Amendment case, Justice Frank Murphy wrote, in a dissenting opinion, that "[f]reedom to think is absolute of its own nature; the most tyrannical government is powerless to control the inward workings of the mind" (Jones v. Opelika, 316 U.S. 584, 618 (1942) (Murphy, J., dissenting)).

The sense of invulnerability individuals have in their interior mental life is undoubtedly connected to the observation I made at the beginning of the book: That an individual is the only person who can have direct experience of her own thoughts and feelings. The rest of the world only has indirect access, and – in the past – often no access at all to the thoughts and memories that she has chosen to refrain from sharing.

I have argued in this book that preserving the security of this space is an interest that should figure powerfully in how constitutional privacy protections are shaped and applied. As proponents of intellectual privacy have argued, surveillance of thought is different – and often far more damaging – than other kinds of privacy invasions. Moreover, while they have focused largely on the way government intrudes into our thinking by accessing reading records, Web searches, and other evidence of our thinking in the outside world, there are good reasons to extend even stronger privacy to our brain operations – and to do so even when they reveal less than many external records of our thinking. In short, we are aware that Web records can be potentially shared, and can still maintain privacy in certain circumstances by retreating into our unshared mental life. But if neuroimaging becomes too powerful, and is left too unconstrained, it may threaten even that refuge. This does not mean that government should be completely barred from obtaining information with this technology. Particularly where individuals voluntarily undergo neuroimaging it may be of benefit to law enforcement, trial fact-finding, and in many other spheres of human life. But protection of our mental privacy requires more robust safeguards than the uncertain and malleable Fourth Amendment limits that the courts have often been quick to pull out of government's way, for example, in warrantless search cases. My argument has been that the First Amendment, therefore, has to be a part of the Fourth Amendment analysis and may even provide a template for an approach that gives weight to mental privacy and autonomy interests in Fifth Amendment self-incrimination law. Not only is the First Amendment relevant because it is the amendment that has often generated judicial commitment to freedom of thought. It is also more capable than Fourth Amendment law, and perhaps Fifth Amendment law, of providing the legal architecture necessary to safeguard the "fortress" or "citadel" that protects our mental life.

REFERENCES

SECONDARY SOURCES

Adelstein, J. S., Shehzad, Z., Mennes, M., DeYoung, C. G., Zuo, X.-N., et al. (2011). Personality is Reflected in the Brain's Intrinsic Functional Architecture. *Plos ONE*, *6*(11), e27633. doi: 10.1371/journal.pone.0027633.

Adkins, D. (2002). The Supreme Court Announces a Fourth Amendment "General Public Use" Standard for Emerging Technologies but Fails To Define It: Kyllo v. United States. *University of Dayton Law Review*, *27*, 245–267.

Adler, M. (1996). Cyberspace, General Searches, and Digital Contraband. *Yale Law Journal*, *105*, 1093–1120.

Alder, K. (2007). *Lie Detectors: The History of an American Obsession*. New York: Simon and Schuster.

Allen, R. J., & Mace, K. M. (2004). The Self-Incrimination Clause Explained and Its Future Predicted. *Journal of Criminal Law and Criminology*, *94*, 243–293.

Amar, A. R., & Lettow, R. B. (1995). Fifth Amendment First Principles: The Self-Incrimination Clause. *Michigan Law Review*, *93*, 857–928.

Ayaz, H., et al. (2011). Suite and Functional Near Infrared Spectroscopy to Study Learning in Spatial Navigation. *Journal of Visualized Experiments*, *56*, e3443. doi: 10.3791/3443.

Bambauer, J. (2015). Other People's Papers. *Texas Law Review*, *94*, 205–263.

Bard, J. S. (2016). Ah, Yes: I Remember It Well: Why the Inherent Unreliability of Human Memory Makes Brain Imaging Technology a Measure of Truth-Telling in the Courtroom. *Oregon Law Review*, *94*, 295–358.

© The Author(s) 2017

129

M.J. Blitz, *Searching Minds by Scanning Brains*,
Palgrave Studies in Law, Neuroscience, and Human Behavior,
DOI 10.1007/978-3-319-50004-1

Battaglia, M., Ogliari, A., Zanoni, A., Citterio, A., Pozzoli, U., Giorda, R., Maffei, C., & Marino, C. (2005). Influence of the Serotonin Transporter Promoter Gene and Shyness on Children's Cerebral Responses to Facial Expressions. *Archives of General Psychiatry, 62*(1), 85–94.

Berlin, I. (1966). *Two Concepts of Liberty.* Oxford, UK: Oxford, Clarendon Press.

Blitz, M. J. (2004). Video Surveillance and the Constitution of Public Space: Fitting the Fourth Amendment to a World That Tracks Image and Identity. *Texas Law Review, 82*, 1349–1481.

Blitz, M. J. (2005). The Dangers of Fighting Terrorism with Technocommunitarianism. *Fordham Urban Law Journal, 32*, 677–719.

Blitz, M. J. (2010). Freedom of Thought for the Extended Mind. *Wisconsin Law Review, 2010*, 1049–1117.

Blitz, M. J. (2016). A Right to Thought Enhancing Technology. In Jotterard, Fabrice & Dubljevic, Veljko (Ed.), *Cognitive Enhancement: Ethical and Policy Implications in International Perspectives.* Oxford, UK: Oxford University Press.

Boire, R. G. (2000, Summer). On Cognitive Liberty, Pt. II. *Journal of Cognitive Liberty, 2*(1), 7–20.

Boire, R. G. (2005). Neurocops: The Politics of Prohibition and the Future of Enforcing Social Policy from Inside the Body. *Journal of Law and Health, 19*, 215–257.

Bonaci, T., & Chizek, H. J. (2013). Privacy by Design in Brain-Computer Interfaces. UWEET Technical Report (pp. 1–3).

Brandom, R. (2015, February 12). Is 'Brain Fingerprinting' a Breakthrough or a Sham? *The Verge.* http://www.theverge.com/2015/2/2/7951549/brain-fin gerprinting-technology-unproven-courtroom-science-farwell-p300.

Brennan-Marquez, K. (2012–13). A Modest Defense of Mind-Reading. *Yale Journal of Law and Technology, 15*, 214–272.

Brown, T., & Murphy, E. (2010). Through a Scanner Darkly: Functional Neuroimaging as Evidence of a Criminal Defendant's Past Mental States. *Stanford Law Review, 62*(1119), 1138–1139.

Buzsaki, G. (2006). *Rhythms of the Brain.* Oxford, UK: Oxford University Press.

Carmel, D., Dayan, E., Naveh, A., Raveh, O., & Ben Shakhar, G. (2003). Estimating the Validity of the Guilty Knowledge Test From Simulated Experiments: The External Validity of Mock Crime Studies. *Journal of Experimental Psychology: Applied, 9*(4), 261–269.

Carraze, A., & Oswald, H. (1996). *The Prisoner: A Televisionary Masterpiece.* London: London Bridge.

Carter, R. (2015, September 25). Neurotelepathy: The Rise of Mind-Reading Machines, Science Focus. http://www.sciencefocus.com/feature/mind-read ing/neurotelepathy-rise-mind-reading-machines.

Chadwick, M. J., et al. (2010). Decoding Individual Episodic Memory Traces in the Human Hippocampus. *Current Biology, 20*(6), 544–547.

Childers, N. (2013, June 6). The Video Game Helmet That Can Hack Your Brain, Motherboard Blog. http://motherboard.vice.com/blog/the-video-game-helmet-that-can-hack-your-brain.

Clark, A., & Chalmers David, J. (2008). The Extended Mind. In A. Clark (Ed.), *Supersizing the Mind: Embodiment, Action, and the Cognitive Experience.* Oxford, UK: Oxford University Press.

Cohen, J. (1996). A Right to Read Anonymously: A Closer Look at "Copyright Management" in Cyberspace. *Connecticut Law Review, 28*, 981–1039.

Committee to Review the Scientific Evidence on the Polygraph, National Research Council, The Polygraph and Lie Detection. (2003). *The Polygraph and Lie Detection.* Washington, DC: National Academies Press.

Cowen, A. S., Chun, M., & Kuhl, B. (2014). Neural Portraits of Perception: Reconstructing Face Images From Evoked Brain Activity. *Neuroimage, 94*, 12–22.

Damasio, A. (2010). *Self Comes to Mind: Constructing the Conscious Brain.* New York: Vintage.

Dick, P. K. (Eds.). (2007). Ubik. In *Four Novels of the 1960s.* New York: Literary Classics of the United States.

Domhoff, G. W. (2003). *The Scientific Study of Dreams: Neural Networks, Cognitive Development and Content Analysis.* New York: American Psychological Association.

Edelman, G., & Tononi, G. (2010). *A Universe of Consciousness: How Matter Becomes Imagination.* New York: Basic Books.

Farah, M. J., Smith, M. E., Gawuga, C., Lindsell, D., & Foster, D. (2010). Brain Imaging and Brain Privacy: A Realistic Concern? *Journal of Cognitive Neuroscience, 21*(1), 119–127.

Farahany, N. A. (2008, April 17). Big Brother Wants to Get in Your Head. Washington Post.

Farahany, N. A. (2012a). Incriminating Thoughts. *Stanford Law Review, 64*, 351–408.

Farahany, N. A. (2012b). Searching Secrets. *Pennsylvania Law Review, 160*, 1239–1308.

Farwell, L. (2012). Brain Fingerprinting: A Comprehensive Tutorial Review of Detection of Concealed Information with Event-Related Brain Potential. *Cognitive Neurodynamics, 6*, 115–154.

Federspiel, W. (2008). 1984 Arrives: Thought(Crime), Technology, and the Constitution. *William & Mary Bill of Rights Journal, 16*, 865–900.

Florence, J., & Friedman, R. (2010). Profiles in Terror: A Legal Framework for the Behavioral Profiling Paradigm. *George Mason Law Review, 17*, 425.

Fox, D. (2008). Will Memory Detection Technologies Transform Criminal Justice in the United States? Brain Imaging and the Bill of Rights. *American Journal of Biethics, 8*(1), 1–4.

Fox, D. (2009). The Right to Silence as Protecting Mental Control. *Akron Law Review, 42,* 763.

Freiwald, S. (2007). First Principles of Communication Privacy. *Stanford Technology Law Review, 2010,* 3–75.

Gamer, M., Klimecki, O., Bauermann, T., Stoeter, P., & Vossel, G. (2012). fMRI-Activation Patterns in the Detection of Concealed Information Rely on Memory-Related Effects. *SCAN, 7,* 506–515.

Gray, D., & Citron, D. (2013). The Right to Quantitative Privacy. *Wisconsin Minnesota Law Review, 98,* 62–144.

Hassabis, D. et al. (2009). Decoding Neuronal Ensembles in the Human Hippocampus. *Current Biology, 19,* 546–554.

Henderson, S. E. (2007). Beyond the (Current) Fourth Amendment: Protecting Third Party Information, Third Parties, and the Rest of Us Too. *Pepperdine Law Review, 34,* 975–1025.

Henderson, S. E. (2011). The Timely Demise of the Fourth Amendment Third Party Doctrine. *Iowa Law Reviews Bulletin, 96,* 39–50.

Holley, B. (2009). It's All in Your Head: Neurotechnological Lie Detection and the Fourth and Fifth Amendments. *Developments in Metal Health Law, 28*(1), 1–23.

Holley, P. (2016, June 5). Their Son Was Killed. They Believe His Parrot Is Telling People Who Pulled the Trigger. *Washington Post.* https://www.washingtonpost.com/news/morning-mix/wp/2016/06/05/their-son-was-killed-they-believe-his-parrot-is-telling-people-who-pulled-the-trigger/?utm_term=.84d253de0f9f

Holloway, M. B. (2008). One Image, One Thousand Incriminating Words. *Temple Journal of Science, Technology & Environmental Law, 27,* 141–174.

Hughes, V. (2014, September 30). The Other Polygraph. *National Geographic: Phenomena.* http://phenomena.nationalgeographic.com/2014/09/30/the-other-polygraph/.

Inception: Warner Bros (2010).

Joh, E. (2006). Reclaiming "Abandoned" DNA: The Fourth Amendment and Genetic Privacy. *Northwestern University Law Reviews, 100,* 858–884.

Jones, O. D., Buckholtz, J. W., Schal, J. D., & Marois, R. (2009). Brain Imaging for Legal Thinkers: A Guide for the Perplexed. *Stanford Technology Law Reviews, 2009,* 1–47.

Kerr, I. R., & McGill, J. (2007). Emanations, Snoop Dogs and Reasonable Expectation of Privacy. *Criminal Law Quarterly, 52*(3), 392–432.

Kerr, O. S. (2009). The Case for Third Party Doctrine. *Michigan Law Review, 107,* 561–601.

Kerr, O. S. (2010). Ex Ante Regulation of Computer Search and Seizure. *Virginia Law Review, 96,* 1241–1293.

Kilbride, M., & Iuliano, J. (2015). Neuro Lie Detection and Mental Privacy. *Maryland Law Review, 75,* 163–193.

Kerr, O. S. (2016, September 9). Thoughts on the Third Circuit's Decryption and Self-incrimination Oral Argument. The Volokh Conspiracy. *Washington Post.*

Kolber, A. (2008). Freedom of memory today. *Neuroethics, 1,* 145–148.

Kolber, A. J. (2016). Two Views of First Amendment Thought Privacy, *University of Pennsylvania Journal of Constitutional Law, 2010,* 1381–1423.

Lacy, J. W., & Stark, E. L. (2013, September). The Neuroscience of Memory: Implications for the Courtroom. *Natural Reviews Neuroscience, 14*(9), 649–658

LaFave, W. R. (1996). *Search and Seizure: A Treatise on the Fourth Amendment* (3rd ed.). St. Paul, MN: Thomson West.

Langleben, D. D., Schroeder, L., Maldjian, J. A., Gur, R. C., McDonald, S., Ragland, J. D., & Childress, A. R. (2002). Brain Activity During Simulated Deception: An Event-Related Functional Magnetic Resonance Study. *NeuroImage, 15,* 727–732.

Langleben, D. T., & Moriarty, J. C. (2013, May 1). Using Brain Imaging for Lie Detection: Where Science, Law, and Policy Collide. *Psychology Public Policy Law, 19*(2), 222–234.

Lee, T. M., Liu, H. L., Tan, L. H., Chan, C. C., Mahankali, S., Feng, C. M., Hou, J., Fox, P. T., & Gao, J. H. (2002, March). Lie Detection by Functional Magnetic Resonance Imaging. *Human Brain Mapping, 15*(3), 157–164.

Levy, N. (2007). *Neuroethics: Challenges for the 21st Century.* Cambridge, UK: Cambridge University Press.

Loftus, E., & Ketchum, K. (1994). *The Myth of Repressed Memory: False Memories and Allegations of Sexual Abuse.* New York: St. Martin's Press.

Loewy, A. H. (1983). The Fourth Amendment as a Device for Protecting the Innocent. *Michigan Law Review, 81,* 1229–1272.

Lykken, D. T. (1959). The GSR in the Detection of Guilt. *Journal of Applied Psychology, 43*(6), 385–388.

Marcuse, L., Fields, M. C., & Yoo, J. (2016). *Rowan's Primer of EEG.* Netherlands: Elsevier.

Martinovic, I., Davies, D., Frank, M., Perito, D., Ros, T., & Song, D. (2012, August). On the Feasibility of Side-Channel Attacks with Brain-Computer Interfaces. In 21st USENIX Security Symposium. USENIX Association.

Moriarty, J. C. (2009). Visions of Deception: Neuroimages and the Search for Truth *Akron Law Review, 42,* 739–760.

Moreno, J. A. (2009). The Future of Neuroimaged Lie Detection and the Law. *Akron Law Review, 42,* 717–736.

Nagel, T. (1974). What is it like to be a bat? *Philosophical Review, LXXXIII*(4), 435–450.

New, J. G. (2008). If You Could Read My Mind. *Journal of Legal Medicine, 29,* 179–198.

Nishimoto, S. et al. (2010). Reconstructing Visual Experiences from Brain Activity Evoked by Natural Movies. *Current Biology, 21,* 1641–1646.

Nozick, R. (1974). *Anarchy, State, and Utopia.* New York: Basic Books.

Ohm, P. (2011). Massive Hard Drives, General Warrants, and the Power of Magistrate Judges. *Virginia Law Reviews in Briefings, 97,* 1–12.

Orwell, G. (1949). *1984.* New York: New American Library.

Pardo, M. S. (2006). Neuroscience Evidence, Legal Culture, and Criminal Procedure. *American Journal of Criminal Law, 33,* 301–337.

Pardo, M. S. (2008). The Self-Incrimination Clause and the Epistemology of Testimony. *Cardozo Law Reviews, 30,* 1023–1045.

Pardo, M. S., & Patterson, D. (2013). *Minds, Brains and Law: The Conceptual Foundations of Law and Neuroscience.* Oxford, UK: Oxford University Press.

PBS NewsHour. (2016, June 28). Solicitor General Donald Verrilli, Who Beat Back Legal Challenges to Obamacare, Steps Down. http://www.pbs.org/news hour/bb/solicitor-general-donald-verrilli-who-beat-back-legal-challenges-to-obamacare-steps-down/.

Pustilnik, A. C. (2013). Neurotechnologies at the Intersection of Criminal Procedure and Constitution Law. In S. Richardson & J. Parry (Eds.), *The Constitution and the Future of Criminal Law.* Cambridge, UK: Cambridge University Press.

Richards, N. (2008). Intellectual Privacy. *Texas Law Review, 87,* 387–445.

Richards, N. (2015). *Intellectual Privacy: Challenges for the 21st Century.* Cambridge, UK: Cambridge University Press.

Rogers, C. (2014). A Slow March Towards Thought Crime: How the Department of Homeland Security's Fast Program Violates the Fourth Amendment. *American University Law Review, 64,* 337–384.

Rowling, J. K. (1998). *Harry Potter and the Chamber of Secrets.* New York: Scholastic.

Sahakian, B. J., & Gottwald, J. (2017). *Sex, Lies, and Brain Scans: How fMRI Reveals What Really Goes on in our Minds.* Oxford, UK: Oxford University Press.

Schauer, Frederick. (2010). Can Bad Science Be Good Evidence? Neuroscience, Lie Detection, and Beyond. *Cornell Law Review, 95,* 1191–1219.

Scherr, A. E. (2013). Genetic Privacy & The Fourth Amendment: Unregulated Surreptitious DNA Harvesting. *Georgia Law Review, 47,* 445–526.

Schulhofer, S. (1991). Some Kind Words for the Privilege Against Self-Incrimination. *Valparaiso University Law Review, 26,* 311–336.

Seung, S. (2012). *Connectome: How the Brain's Wiring Makes Us Who We Are.* New York: Houghton Mifflin.

Shen, F. X. (2013). Neuroscience, Mental Privacy and the Law. *Harvard Journal of Law and Public Policy, 36,* 653–713.

Shen, F. X. (2016). Neurolegislation: How U.S. Legislators Are Using Brain Science. *Harvard Journal of Law & Technology, 29*, 495–526.

Shinkareva, S. V. et al. (2008). Using fMRI Brain Activation to Identify Cognitive States Associated with Perception of Tools and Dwellings. *Plos ONE, 3*(1), e1394. doi: 10.1371/journal.pone.0001394.

Slobogin, C. (2007). *Privacy at Risk: The New Government Surveillance and the Fourth Amendment.* Chicago: University of Chicago Press.

Slobogin, C. (2014). Panvasive Surveillance, Political Process Theory, and Non-Delegation Doctrine. *Georgetown Law Journal, 102*, 1721–1775.

Smith, K. (2013, October 13). By Scanning People's Brain Activity, Scientists May Be Able to Decode People's Thoughts, Their Dreams, Even Their Intentions. *Nature, 502*, 428–430.

Snead, O. C. (2007). Neuroimaging and the "Complexity" of Capital Punishment. *New York University Law Review, 82*, 1265–1339.

Solove, D. J. (2007). First Amendment as Criminal Procedure. *New York University Law Reviews, 82*, 112–176.

Solove, D. J. (2010). Fourth Amendment Pragmatism. *Boston College Law Review, 51*, 1511–1538.

Spence, S. (2004). A Cognitive Neurobiological Account of Deception: Evidence from Functional Neuroimaging. *Philosophical Transactions of the Royal Society B: Biological Sciences, 359*(1451), 1755–1762. PMC. Web. 5 Sept. 2016.

Stoller, S. E., & Wolpe, P. R. (2007). Emerging Technologies for Lie Detection and the Fifth Amendment. *American Journal of Law and Medicine, 33*(2/3), 359–374.

Stromberg, J. (2013, April 4). Scientists Figure Out What You See While You're Dreaming. http://Smithsonian.com http://www.smithsonianmag.com/science-nature/scientists-figure-out-what-you-see-while-youre-dreaming-15553304/.

Stuntz, W. J. (1988). Self-Incrimination and Excuse. *Columbia Law Review, 88*, 1227–1296.

Stuntz, W. J. (1995). Privacy's Problem and the Law of Criminal Procedure. *Michigan Law Review, 93*, 1016–1078.

Sur, S., & Sinha, V. K. (2009, Jan–Jun). Event Related Potential: An Overview. *Industrial Psychiatry Journal, 18*(1), 70–73.

Strange Days: Lightstorm Entertainment/20th Century Fox (1995).

Thompson, S. K. (2007). A Brave New World of Interrogation Jurisprudence. *American Journal of Law & Medicine, 33*, 341–357.

Tovino, S. A. (2005). The Confidentiality and Privacy Implications of Functional Magnetic Resonance Imaging. *Journal of Law and Medical Ethics, 33*, 844–848.

Village of the Damned, Metro-Goldwyn Mayer (1960).

Ward, J. (2010). *Student's Guide to Cognitive Neuroscience* (2nd edn.). New York: Psychology Press.

Tovino, S. A. (2007). Functional Neuroimaging Information: A Case of Neuroexceptionalism. *Florida State University Law Review, 33*, 415–489.

Tribe, L. H. (1988). *American Constitutional Law* (2nd edn.; 1321–1326). St. Paul, MN: Foundation Press.

Wagner A., et al. (2016). fMRI and Lie Detection: A Knowledge Brief of the MacArthur Foundation Research Network on Law and Neuroscience. http://www.lawneuro.org/LieDetect.pdf

Winick, B. J. (1989). The Right to Refuse Mental Health Treatment: A First Amendment Perspective. *University of Miami Law Review, 44*, 1–103.

Wolpaw, J., & Wolpaw, E. W. (2012). *Brain-Computer Interfaces: Principles and Practice*. Oxford, UK: Oxford University Press.

Yuhas, D. (2012, June 12). What's a Voxel and What Can It Tell Us? A Primer on fMR. *Scientific American Blog*. http://blogs.scientificamerican.com/observations/whats-a-voxel-and-what-can-it-tell-us-a-primer-on-fmri/.

INDEX

© The Author(s) 2017
M.J. Blitz, *Searching Minds by Scanning Brains*,
Palgrave Studies in Law, Neuroscience, and Human Behavior,
DOI 10.1007/978-3-319-50004-1

Name Index

© The Author(s) 2017   141
M.J. Blitz, *Searching Minds by Scanning Brains*,
Palgrave Studies in Law, Neuroscience, and Human Behavior,
DOI 10.1007/978-3-319-50004-1

COURT CASES

© The Author(s) 2017

M.J. Blitz, *Searching Minds by Scanning Brains*, Palgrave Studies in Law, Neuroscience, and Human Behavior, DOI 10.1007/978-3-319-50004-1